T0312153

Routledge Studies in Environmental Policy

Strategic Designs for Climate Policy Instrumentation
Governance at the Crossroads
Gjalt Huppes

The Right to Nature
Social Movements, Environmental Justice and Neoliberal Natures
Edited by Elia Apostolopoulou and Jose A. Cortes-Vazquez

Guanxi and Local Green Development in China
The Role of Entrepreneurs and Local Leaders, 1st Edition
Chunhong Sheng

Environmental Policy in India
*Edited by Natalia Ciecierska-Holmes, Kirsten Jörgensen, Lana Ollier
and D. Raghunandan*

Mainstreaming Solar Energy in Small, Tropical Islands
Cultural and Policy Implications
Kiron C. Neale

EU Environmental Governance
Current and Future Challenges
Edited by Amandine Orsini and Elena Kavvatha

The European Union and Global Environmental Protection
Transforming Influence into Action
Edited by Mar Campins Eritja

Environmental Policy and Air Pollution in China
Governance and Strategy
Yuan Xu

For more information about this series, please visit: www.routledge.com/
Routledge-Studies-in-Environmental-Policy/book-series/RSEP

Environmental Policy and Air Pollution in China

Governance and Strategy

Yuan Xu

LONDON AND NEW YORK

First published 2021
by Routledge
2 Park Square, Milton Park, Abingdon, Oxon OX14 4RN

and by Routledge
52 Vanderbilt Avenue, New York, NY 10017

Routledge is an imprint of the Taylor & Francis Group, an informa business

© 2021 Yuan Xu

The right of Yuan Xu to be identified as author of this work has been
asserted by him in accordance with sections 77 and 78 of the Copyright,
Designs and Patents Act 1988.

The Open Access version of this book, available at www.taylorfrancis.
com, has been made available under a Creative Commons Attribution-Non
Commercial-No Derivatives 4.0 license.

Trademark notice: Product or corporate names may be trademarks or
registered trademarks, and are used only for identification and explanation
without intent to infringe.

British Library Cataloguing-in-Publication Data
A catalogue record for this book is available from the British Library

Library of Congress Cataloging-in-Publication Data
A catalog record for this book has been requested

ISBN: 978-1-138-32232-5 (hbk)
ISBN: 978-0-429-45215-4 (ebk)

Typeset in Times New Roman
by Apex CoVantage, LLC

Contents

Figures

Tables

Preface

China is puzzling to read.

After the Cultural Revolution and a short transitional period, China entered the era of Reform and Open-up in December 1978. The size of China's economy has skyrocketed by more than 30 times. Despite numerous benefits, this rapid economic growth also brought immense pressure on the environment. China's environmental crises are multifaceted, stretching across air, water, soil, ecosystem and climate change.

Hope was not readily available. As a public good, environmental protection requires effective governmental intervention. However, China is not a democracy, and sound rule of law has not been established. The country's governance quality has been ranked consistently and significantly lower than that in developed countries that are liberal democracies, where environmental quality first deteriorated with economic growth and then fundamentally improved. Their experiences suggest that China's environmental crises are expected, while their solutions are hard to reach.

Then what happened in China in the past 15 years became surprising as the environmental trajectory deviated away from the projections. Sulfur dioxide (SO_2) is one air pollutant that is crucial for air quality but very difficult to control. Since reaching their peak in the mid-2000s, SO_2 emissions in China have been declining, and the downward pace accelerated in the past few years to reach a level not seen in more than four decades. A large coal-fired power sector appeared to install and operate SO_2 scrubbers that mitigate emissions from polluting sources. Similar desirable outcomes are also observed in other environmental and renewable energy fields. However, China has not changed seriously from the perspectives of democracy and the rule of law, although environmental policy has been improving and strengthening. The legal system still does not play any major role in environmental protection. Policy making lacks transparency and public consultation, while policy blunders are not rare. Policy implementation still has considerable problems and is often selective. It is not unusual to hear about the abuse of governmental authorities.

This book aims to provide a theoretical understanding to explain how China achieved deep and sustained pollution mitigation without democracy and sound rule of law. Causal relationships are explored between the favorable outcome and the unfavorable path. The major puzzle is why China frequently witnesses both sides at the same time or whether the conventional insights may have missed something important in reading China. China's strategy is theorized into *goal-centered governance*. China is both highly centralized – in goal setting – and highly decentralized – in goal attainment, policy making and implementation. Unlike the rule-based governance in developed countries as indicated in their well-established rule of law, China places goals in the first place, while deficiencies in policy making and implementation are much tolerated as long as goals can be attained. The mitigation trajectory was not centrally planned but gradually evolved through decentralized pathfinding under centralized goals. In other words, the Chinese puzzle should primarily be explained from the perspective of its governance strategy but not individual policies. A strategic mistake is often a lot more devastating and far-reaching than any policy stumble, while an effective strategy can accommodate many policy mistakes without compromising much the final outcome.

The research and thinking for this book stretched over a dozen years. When I first started studying China's SO_2 mitigation around 2007, the hypothesis was that the environmental crisis was rooted in policy failures and, more fundamentally, the lack of democracy and the rule of law. However, what unfolded later forced me to rethink this causal relationship, especially in the 2010s when the mitigation pace dashed forward. As a former physicist, I hope to find a theoretical explanation to the Chinese puzzle that is simple, like one equation, and rich. The *goal-centered governance* in this book reflects such a new attempt.

Acknowledgments

I owe a tremendous amount of debts to many people. This book is dedicated to Robert H. Socolow, the supervisor of my PhD thesis at Princeton University's Woodrow Wilson School of Public and International Affairs. His inspiration is vital in my research journey. Much of this book is rooted although widely extended from my PhD study over a decade ago. I am grateful for Robert H. Williams, Denise L. Mauzerall, Eric D. Larson, Yiguang Ju, Gregory C. Chow, Edward S. Steinfeld, Richard K. Lester and Kin-Che Lam, whose support and insights were crucial to sustain and enlighten this research. My deep appreciation also goes to numerous interviewees who kindly shared their knowledge. I thank Matthew Shobbrook of Routledge, whom I worked with to finally complete this book.

My wife, Jing Song, and our two children, Anlan Xu and Antao Song, are perpetual motivation and sources of encouragement for my research. My parents, Meilan Yuan and Yicai Xu, and parents-in-law, Meiyu Song and Changfa Song, provide patient and unconditional support. My family made this work possible, especially under the ongoing COVID-19 pandemic.

Funding support throughout this research in the past dozen years was provided by Princeton University, Massachusetts Institute of Technology, The Chinese University of Hong Kong, and Hong Kong Research Grants Council (General Research Fund, 14654016).

Parts of the book were adapted with permissions from the author's several published journal articles, including Xu, Y. 2011. The use of a goal for SO_2 mitigation planning and management in China's 11th five-year plan. *Journal of Environmental Planning and Management*, 54, 769–783 [in Chapter 4; Copyright (2011) Taylor & Francis]; Xu, Y. 2011. Improvements in the operation of SO_2 scrubbers in China's coal power plants. *Environmental Science & Technology*, 45, 380–385 [in Chapter 6; Copyright (2011) American Chemical Society]; Xu, Y. 2011. China's functioning market for sulfur dioxide scrubbing technologies. *Environmental Science & Technology*, 45, 9161–9167 [in Chapter 7; Copyright (2011) American Chemical Society]; Xu, Y. 2013. Comparative advantage strategy for rapid pollution mitigation in China. *Environmental Science & Technology*, 47, 9596–9603 [in Chapter 7; Copyright (2013) American Chemical Society]. Much has been revised and expanded on.

1 Introduction

1 China's environmental crises

China faces colossal, multifaceted environmental challenges, many at crisis levels. Its environmental degradation has been widely documented and analyzed in academic studies as well as in public media. China is now the largest energy consumer, supplier and emitter of most major air and water pollutants as well as various greenhouse gases. Together with its geographically high population and economic densities, especially in the eastern half of the country, China was categorized at the very bottom of air quality among the 180 countries and regions in the Environmental Performance Index (Wendling et al., 2018; Figure 1.1). Few readers would be surprised to know that China's air quality is among the most polluted in the world (Figure 1.1).

Air and water pollution in China have certainly taken a serious toll. China has made steady progress in the past decades to significantly reduce premature deaths due to water-related environmental factors and indoor air pollution, but ambient particulate matter (PM) pollution has been deteriorating. The Global Burden of Disease study elaborates in great detail the causes and risk factors of deaths across individual countries (Institute for Health Metrics and Evaluation, 2018). In 1990, China accounted for 22.2% of the global population, and in 2017, the share dropped to 18.5% despite an 18.0% increase in absolute population (Figure 1.2). In premature deaths that are due to environmental risk factors, China's share in the world in 1990 was 29.2% for household air pollution from solid fuels and 4.6% for unsafe water, sanitation and handwashing. In other words, an average Chinese was 31.5% more likely and 79.3% less likely to die prematurely due to the two risks than an average person in the world. The shares were significantly reduced to 16.5% and 0.6% in 2017, respectively, to make an average Chinese 10.6% and 96.8% less likely to die prematurely. In absolute terms, they were reduced by 65.7% and 92.5%, respectively. However, ambient PM pollution caused 404,000 premature deaths in 1990 and 852,000 in 2017, more than double. Its global share climbed from 23.0% to 29.0% over the period. In 2000, indoor air pollution was overtaken by ambient PM pollution in causing more premature deaths. In comparison to China's share of the global population, in 1990, an average Chinese faced only a slightly greater risk, 3.8%, from ambient PM pollution than an average person in the world, but in 2017, the risk premium was enlarged to 56.9%.

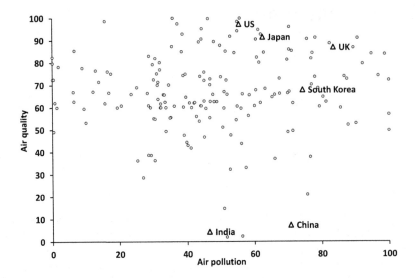

Figure 1.1 Environmental Performance Index in the baseline year

Source: Wendling et al. (2018).

Note: "Air pollution" at the *x*-axis refers to sulfur dioxide (SO_2) and nitrogen oxide (NO_x) emission inten-sities, and its baseline year is 2006. "Air quality" in the *y*-axis indicates household solid fuels (baseline year: 2005), fine particulate matter ($PM_{2.5}$) exposure and $PM_{2.5}$ exceedance (baseline year: 2008).

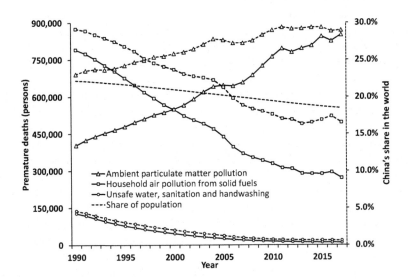

Figure 1.2 China's premature deaths due to air and water pollution in the Global Burden of Disease study

Source: Institute for Health Metrics and Evaluation (2018).

Note: Solid lines indicate absolute numbers in persons with the left *y*-axis, while dashed lines refer to China's shares in the world with the right *y*-axis.

Another measurement of pollution's health impact is the disability-adjusted life years (DALYs) that quantifies the loss of "healthy" life years. It combines the lost life years due to both premature deaths and illnesses. Various types of environmental pollution in different countries may cause premature deaths and illnesses that correspond to different life expectancies, ages and other situations. The ratio between DALYs and premature deaths is much higher for water pollution than for air pollution. For example, in 2017, China lost 19.8 million, 6.46 million and 0.85 million DALYs due to ambient PM pollution, household air pollution from solid fuels, and unsafe water, sanitation and handwashing, respectively. The corresponding ratios between DALYs and premature deaths were 23.3, 23.8 and 89.0, respectively, to indicate the more severe health impacts of water pollution for an average case.

Nevertheless, the indicator of DALYs does not change the conclusion that was presented with the examination of premature deaths (Figure 1.3). Substantial progress was also made on indoor air pollution and water, with their DALYs being reduced by 77.3% and 92.0%, while the deterioration trend for ambient PM pollution is distinguished with an increase of DALYs by 47.3%. In terms of China's shares in the world, ambient PM pollution is still the only risk factor among the three to surpass that of its population, which accounted for 23.8% of the world's total in 2017. For all DALYs due to the three environmental risk factors, ambient PM pollution's share rose from 25.6% in 1990 to 73.0% in 2017. Accordingly,

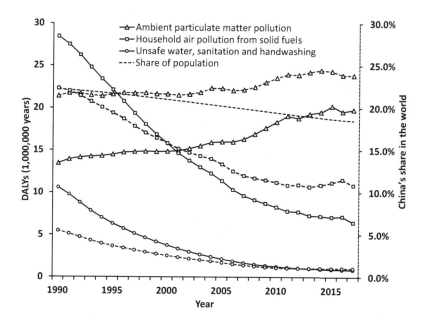

Figure 1.3 Disability-adjusted life years (DALYs) in China due to air and water pollution in the Global Burden of Disease study

Source: Institute for Health Metrics and Evaluation (2018).

environmental pollution in China is more and more dominated by ambient air pollution and especially PM pollution.

On average, the DALYs due to various environmental risks indicate that an average Chinese loses a significant number of healthy life days for every year living in these environmental risks. In 1990, household air pollution from solid fuels was the most severe environmental risk in China to incur the loss of 8.7 disability-adjusted life days (DALDs) per person, while the damages from ambient PM pollution and from unsafe water, sanitation and handwashing were similar at 4.1 and 3.2 DALDs per person, respectively (Figure 1.4). In other words, an average Chinese lost 16.0 health life days due to the three air and water pollution risk factors for living through 1990. In 2017, ambient PM pollution became the most severe risk factor, being responsible for 5.1 DALDs per person or 1.0 DALDs more, after the other two experienced dramatic improvement in the past decades. The total loss was 7.0 DALDs for living through 2017.

China is not a unique country to witness the diverging progress of different risk factors. India had similar paths for distinguishing the rising importance of ambient PM pollution in environmental protection. Ambient PM pollution in India has remained stable throughout the years to account for 5.7 and 5.6 DALDs per person in 1990 and 2017, respectively. Although household air pollution from solid fuels still claimed greater health damages in 2017, its steady declining trend

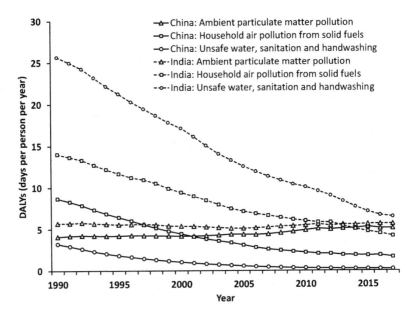

Figure 1.4 DALYs in days (or disability-adjusted life days [DALDs]) per person per year in China and India

Source: Institute for Health Metrics and Evaluation (2018).

suggests that ambient PM pollution will soon become the most damaging environmental risk among the three in India as well (Figure 1.4).

2 China's expected rise of SO$_2$ emissions and unexpected success in SO$_2$ mitigation

China has been rapidly industrializing in the past four decades. Environmental crises can be empirically expected in the contexts of its rapid economic development, rising energy consumption and coal dominance. The expectation also comes from crucial governance factors that are believed to be favorable for environmental transition but that China is especially weak at. First, democracy is believed to be good for environmental protection by many scholars (e.g., Payne, 1995). Unfortunately, China is not a democracy, and thus, society's demand for cleaner air is often not believed to be able to effectively influence policy making as in a democracy. It is generally ranked at the bottom of various democracy indexes. According to Polity's ratings that can reflect the common views of democracy evaluation at least in Western liberal democracies, modern-day China, under the communist rule, is debatably less democratic than the imperial days in the 19th-century Qing dynasty, when the emperors still held absolute power, with the Polity index being −6 (Marshall et al., 2019). China's economic reform era after the Cultural Revolution only slightly

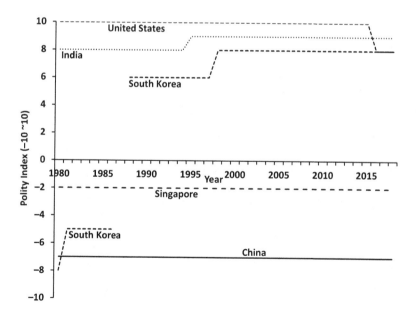

Figure 1.5 Polity Democracy Index for China, South Korea, Singapore, India and the United States (−10 being the most autocratic and 10 the most democratic)

Source: Marshall et al. (2019).

improved its Polity index from -8 to -7 (Figure 1.5). In comparison, South Korea was fundamentally transformed from an authoritarian regime to a democratic one after the reform in the 1980s. Singapore is steadily ranked toward the authoritarian side. India and the United States are standard democracies despite slight fluctuations.

Democratic states are argued to be more responsive to the public's demands. If the public in a democracy gives top priority to environmental matters, strong political will is more likely to be generated (Li and Reuveny, 2006; Payne, 1995; Downey and Strife, 2010). Furthermore, the public in a democracy could be more pro-environment than are the elites in an autocracy; this could be because of better access to information, a more developed civil society and a longer time horizon of planning (Li and Reuveny, 2006; Payne, 1995). Democracy is generally closely associated with the rule of law, and therefore, there should be better enforcement of environmental regulations (Li and Reuveny, 2006). Nevertheless, democracy might also be associated with weakness in environmental protection. People's self-interest and the interests of business are more difficult to overcome in a democracy (Li and Reuveny, 2006). If the public gives only a low priority to having a clean environment, then a democracy could be less likely to heavily focus on environmental protection.

Empirical statistical studies have found no conclusive relationship between democracy and the environment. Congleton (1992) and Neumayer (2002) found that democracy contributes positively to international environmental commitments. Midlarsky (1998) discovered that democracy leads to more protected areas of land, but that it tends to negatively influence deforestation and carbon dioxide (CO_2) emissions per capita. Winslow (2005) found only good effects of democracy, whereas Pellegrini and Gerlagh (2006) found that it had insignificant impacts. The mixed results of the relationship could be at least partly caused by the difference in environmental indicators. For example, CO_2 is more difficult to abate, but it has much less local influence than urban particulate pollution. Studies that used panel data also reported mixed results regarding the relationship (Torras and Boyce, 1998; Barrett and Graddy, 2000). Different democracy indexes do not differ greatly in their relationship to the environment. A problem in the literature is that a linear relationship is generally assumed between democracy and the environment. However, theoretical arguments might suggest that both democracy and autocracy could have a beneficial effect on environmental protection, while regimes in between make the situation worse. Among control variables, the most common one is income. Considering the literature on the Environmental Kuznets Curve and a plausible relationship between income and the environment (Grossman and Krueger, 1995; Stern and Common, 2001), income together with its squared and cubed terms are necessary control variables. One study that did not include income as an independent variable could suffer from potential missing-variable problems (Winslow, 2005). In addition, two studies controlled a governance index, namely, that of corruption (Pellegrini and Gerlagh, 2006; Buitenzorgy and Mol, 2011), but most of them disregarded governance. Various studies differ greatly from each other in how they control other variables, including trade openness (Li and Reuveny, 2006), inequality/ Gini ratio (Torras and Boyce, 1998), energy resource endowment (Congleton,

1992), country size in gross domestic product (GDP; Winslow, 2005), population size (Neumayer, 2002; Congleton, 1992) and literacy (Torras and Boyce, 1998).

Case studies found no conclusive relationship either. A case study in Kenya found that democracy is benign to the environment; this is because the government responded mainly to the "environmental and developmental civil society" and "Western supporters" rather than to the "marginalized poor" (Njeru, 2010). On the other hand, democratization in a number of southern African countries, particularly Malawi, South Africa and Mozambique, has resulted in greater destruction of the environment for short-term economic and social reasons (Walker, 1999). In Mexico City, it has been found that democratic elections do not assist in stopping local deforestation (Hagene, 2010). Through studying China and Southeast Asia, it is even proposed that " 'good' authoritarianism" is essential for solving our urgent environmental problems (Beeson, 2010). A case study in Guatemala found that the relationship between democracy and the environment is complex and not straightforward (Sundberg, 2003).

In addition, the empirical relationship between economic development and environmental quality did not expect that China would be able, or willing, to pull down its pollutant emissions and improve air quality. Environmental Kuznets Curve – an empirical bell-shaped relationship between income level and environmental quality – predicts that before a country becomes rich enough to reach a certain level of income (or GDP per capita), its environmental quality will keep

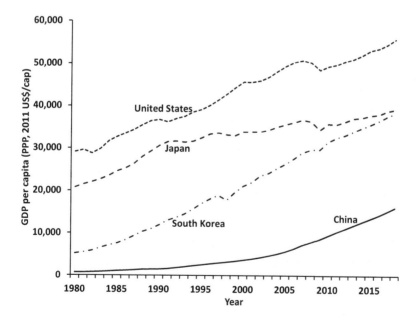

Figure 1.6 GDP per capita in PPP (purchasing power parity) in China, South Korea, Japan and the United States

Source: IMF (2019).

deteriorating (Grossman and Krueger, 1995). China's GDP per capita in purchasing power parity and constant 2011 dollars in 2018 was US$16,100, and the International Monetary Fund projected that it would rise to US$22,200 in 2024, while the level was US$29,100 in the United States in 1980 (Figure 1.6). In other words, China is about five decades behind the United States in terms of economic development status. Different studies report different turning points, and the lowest one for SO_2 emissions is at about US$3,000 (in 1990 US$ and nominal exchange rates) (Stern and Common, 2001). China's GDP per capita only surpassed US$3,000 per capita in nominal terms in 2008 (IMF, 2019), which was still much lower than the empirical minimum turning point.

Furthermore, environmental governance is critical to provide better environmental quality as a public good. As suggested in the World Bank's six governance indicators, comparatively China is poorly governed (Kaufmann and Kraay, 2019). The indicators assigned a score between −2.5 (worst) and 2.5 (best) to indicate governance performance. On "voice and accountability," China scored consistently and significantly lower than democracies, such as the United States and India. Their average scores from 1996 to 2018 were −1.58, 1.18 and 0.42, respectively

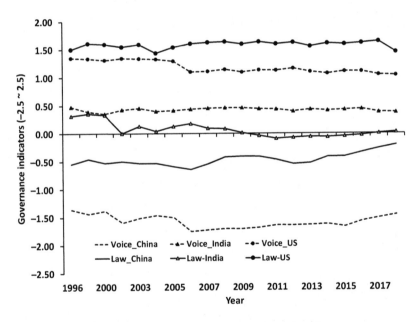

Figure 1.7 Governance indicators of China, India and the United States

Source: Kaufmann and Kraay (2019).

Note: "Voice": *Voice and accountability* "reflects perceptions of the extent to which a country's citizens are able to participate in selecting their government, as well as freedom of expression, freedom of association, and a free media." "Law": *Rule of law* measures "perceptions of the extent to which agents have confidence in and abide by the rules of society, and in particular the quality of contract enforcement, property rights, the police, and the courts, as well as the likelihood of crime and violence."

(Kaufmann and Kraay, 2019; Figure 1.7). It suggests that Chinese citizens are less able to directly participate in selecting a government and that their voices are less likely to be heard. In terms of "political stability and absence of violence/terrorism," China scored −0.44, better than India's −1.13 but worse than United States' 0.48. "Government effectiveness" measures the provision of public and civil services as well as the quality of policy making and implementation. It is the governance indicator that China had the best performance. It is also the only one that China's score is positive, being 0.09 on average, and consistently improved from −0.35 in 1996 to 0.48 in 2018 (Kaufmann and Kraay, 2019). Nevertheless, China is still much behind the United States that scored 1.58 in 2018. For the "rule of law" indicator, China performs poorly with an average score of −0.46, much lower than the United States' 1.58 and India's 0.07 (Figure 1.7). Although slight progress was made in China from −0.55 in 1996 to −0.20 in 2018, it was always located in the negative territory. Little progress was achieved on "corruption" as the score remained consistently low with an average of −0.41, which was poorer than the 1.47 in the United States and −0.38 in India (Kaufmann and Kraay, 2019). China performed steadily poor in "regulatory quality" that focuses on the private sector. The United States scored 1.51 on average for the 1996–2018 period, much better than China's −0.25 or India's −0.36 (Kaufmann and Kraay, 2019). These governance indicators quantitatively measure various aspects of governance in a country to enable comparison across countries and years. As a classical example of market failure to demand governmental intervention, environmental protection cannot be effective without effective governance. However, none of the six governance indicators suggest that the Chinese government can sustainably, effectively and efficiently enact and implement environmental policies and laws.

With all the unfavorable conditions and rising environmental pressures from energy consumption, little hope existed to make China's environmental cleanup promising. SO_2 is one of the most important air pollutants, and it was also the first air pollutant explicitly included in the national Five-Year Plans for serious mitigation (National People's Congress, 2006). Its emissions were more than doubled from 1980 to the 2000s to echo such expectations (Figure 1.8). However, something has obviously worked as indicated in the more recent trajectory of SO_2 emissions (Figure 1.8). Multiple data sources – from Chinese official statistics, independent bottom-up and top-down estimates inside and outside of the country to satellite and remote sensing data – all point to the same trend: China's SO_2 emissions have been rapidly decreasing in the past decade (Li et al., 2017; Zheng et al., 2018; Lu et al., 2011; Crippa et al., 2018; Fioletov et al., 2019; National Statistics Bureau and Ministry of Ecology and Environment, 2019). Although different emission inventories still show gaps between each other on when peak SO_2 emissions happened and how high they reached, China should have completely wiped out all additional SO_2 emissions that accompanied its unprecedented economic growth in the past four decades (Figure 1.8). Although China's economy has expanded by more than 30-fold since the Open-up policy was initiated in 1978, the country now emits significantly less SO_2 (Figure 1.8). It seems to have taken China less than one decade to remove all the additional

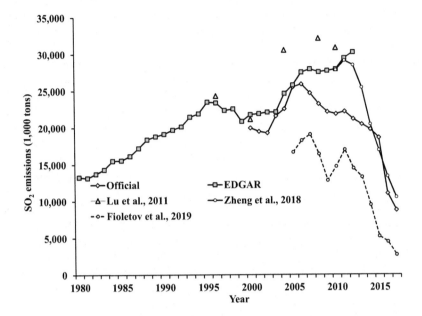

Figure 1.8 SO$_2$ emissions in China

Source: Data from Fioletov et al. (2019) refer to large power plants, while others are for China as a whole (Zheng et al., 2018; Lu et al., 2011; Crippa et al., 2018; Fioletov et al., 2019; National Statistics Bureau and Ministry of Ecology and Environment, 2019).

SO$_2$ emissions that the country increased with its economic development and energy consumption.

With the rapid electrification trend of energy consumption and the power sector's increasing share of coal consumption, the power sector is becoming more and more important in deciding the trajectory of China's SO$_2$ mitigation. In 1980, its share of SO$_2$ emissions was only 22.5%, less than the industrial sector's 50.0% and the residential sector's 23.5% (Figure 1.9). The relatively less significance was due to the power sector's low share of coal consumption, 20.2% (Figure 1.10). In the following two decades, the power sector's share climbed continuously to peak in 2002 at 45.7% and surpass that of the industrial and residential sectors (Figure 1.9) together with its 52.2% share of coal consumption (Figure 1.10). However, these two trajectories started to diverge from each other afterward (Figure 1.10). In 2017, the power sector consumed 57.3% of China's coal but only accounted for 17.4% of SO$_2$ emissions (Figure 1.9). The industrial and residential sectors' shares rebounded to reach 56.8% and 22.6%, respectively. Accordingly, the power sector now emits much less SO$_2$ for consuming one unit of coal than the industrial and residential sectors do.

Although energy transition away from coal is favorable for SO$_2$ mitigation, coal consumption in China still remains at a high level, with only a slight decrease

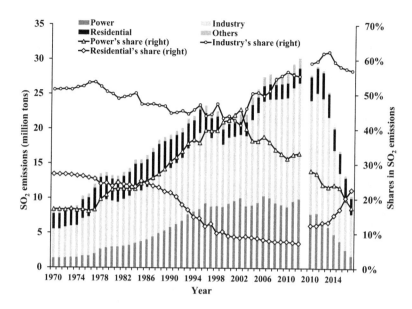

Figure 1.9 SO$_2$ emissions by sector in China (from two different data sources for 1970–2012 and 2010–2017, respectively)

Source: Crippa et al. (2018); Zheng et al. (2018).

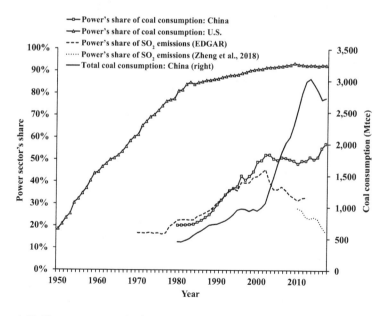

Figure 1.10 The power sector's shares of coal consumption and SO$_2$ emissions in China and the United States

Source: EIA (2019); Fridley and Lu (2016); National Bureau of Statistics (2019).

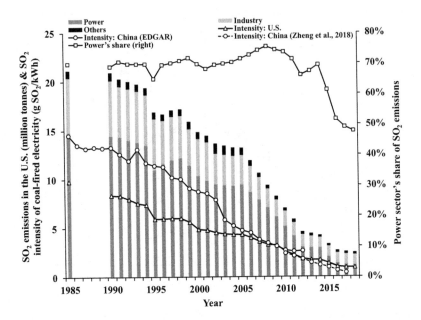

Figure 1.11 SO$_2$ emissions in the United States and SO$_2$ intensities in China and the United States

Source: Crippa et al. (2018); Zheng et al. (2018); BP (2019); U.S. EPA (2019).

in recent years (Figure 1.10). Most mitigation of absolute SO$_2$ emissions was because much greater SO$_2$ emissions are avoided per unit of coal consumption. The power sector's high efficiency in removing SO$_2$ is also reflected in its SO$_2$ emission intensity of coal-fired electricity. Since the enactment of the Clean Air Act Amendments (1990), the United States has substantially reduced its overall SO$_2$ emissions from 20.9 million tons in 1990 to 2.48 million tons in 2018 (Figure 1.11). The power sector has consistently been the largest contributor, and its SO$_2$ emissions dropped from 14.4 million tons to 1.19 million tons over the same period, while its share declined from 68.9% to 47.8%. The much higher share than China's reflects the power sector's greater importance in U.S. coal consumption (Figure 1.10).

In reference to the successful progress in the United States, China's SO$_2$ mitigation trajectory was even steeper. In 1990, for generating 1 kWh of coal-fired electricity, 8.4 g of SO$_2$ were emitted in the United States, while the rate was 58.5% higher, or 13.3 g in China. In 2017, as calculated with independent emission inventory data, the SO$_2$ intensity decreased to be 0.96 g in the United States and 0.41 g in China, 56.9% lower (Figure 1.11).

3 The organization of this book

Democracy and rule of law have played prominent and indispensable roles in environmental cleanup in developed countries. However, China is not a democracy and political freedom is indeed highly constrained, but why environmental protection became the country's priority to witness a dramatic drop in SO_2 emissions? Furthermore, many policies are not implemented well, and the legal system plays an essentially negligible role in China's environmental protection. But why the government was able to effectively bend down pollutant emissions at such an astonishing pace? This book focuses on how China defied the empirical expectations in SO_2 mitigation, especially in the coal-fired power sector. It aims to provide an explanation at the strategic level for understanding how environmental governance is organized and implemented in China.

This book also aims to imply China's governance in general. Observers on China's governance often have polarized views and each side seems to have ample supporting evidence. Regardless of what the focused perspective is, China is full of puzzles and controversies. The country has made many remarkable achievements in the past 40 years, with much higher income and living standards, much better infrastructures, much wider social safety nets, much less control of individuals' private lives and much less poverty. It leads the world in renewable energy development and electric vehicles. However, rules are much less respected in China than in developed countries. The parliament – the National People's Congress – is often referred to as a "rubber stamp," although in the Chinese Constitution, it has the utmost authority beyond any governmental entity. The judicial system is not independent. Political liberty is much constrained without genuine elections. The Chinese Communist Party has almost unchecked power, and the authoritarian country is ruled from the top, but an often-heard sentence in China goes that "policies and orders cannot go beyond *Zhongnanhai*" (the compound where the central government is located). How should we explain China's governance and reconcile the polarized observations that are both well documented and evidence-based? Are the two sides causally connected? How can China achieve those favorable outcomes with such an unfavorable policy pathway? If we repair all recognized deficiencies in the governance, are we going to throw away the baby together with the bathwater? Most important, does China follow a different governance model from that in developed countries, and thus, is the explanatory power of many theories and historical experiences reduced?

The rest of the book is organized as follows: Chapters 2, 3 and 4 examine how the Chinese government is organized for environmental protection, especially in the contexts of neither democracy nor sound rule of law. Chapter 2 explores how the political will for environmental protection has been centrally evolving without democracy. Chapter 3 discusses China's environmental governance structure that combines high degrees of both centralization and decentralization from different perspectives. Primary focuses are on the evolution of the Ministry of Ecology and Environment and the relationships between the central and local governments. Chapter 4 studies how prioritized environmental protection is transmitted from the central government to local governments for their effective mobilization against the background of a weak rule of law. The environmental governance is organized to center on goals, specifically on SO_2 emissions and environmental protection in Five-Year Plans. This book calls the governance strategy in China as the goal-centered governance model that features centralized goal setting and decentralized goal attainment.

Chapters 5, 6 and 7 analyze the impacts of China's goal-centered governance model. Chapter 5 focuses on decentralized policy making for SO_2 mitigation that is guided by centralized, top-down goals. This integration of centralization and decentralization has generated not only profound outcomes, with active policy making, innovation and competition, but also many policy deficiencies. China's governance is tolerant of mistakes or even abuses in policy making, as long as goals can be achieved. Such tolerance then significantly reduces the requirements for policy making quality, choices of policy instruments and inter-policy coordination. Chapter 6 explores how this goal-centered governance has exerted impacts on decentralized policy implementation. From unfavorable backgrounds of inadequate capacity, effectiveness and efficiency of environmental policy implementation, local governments make gradual and steady improvements that aim for approaching their assigned goals. Chapter 7 addresses how China overcame supply constraints and established its domestic SO_2 scrubber industry for meeting the skyrocketing demand. Decentralized market entities were able to actively seek and capture market opportunities under goal-centered governance. Goals on environmental protection and economic development could thus achieve better synergies than conflicts.

Chapter 8 concludes this book and discusses the goal-centered governance model. This theoretical framework can integrate the polarized observations on China within a systematic and compatible understanding. The rule-based governance model is the primarily applied strategy in countries with sound rule of law that emphasizes on making good, often centralized policies as means, but the final outcome is less explicit. In contrast, this goal-centered governance model emphasizes centralized goals as ends but is more relaxed on the means to result in many policy deficiencies. In the contexts of China's backgrounds of no democracy and weak rule of law, this governance strategy has been proved effective not only on SO_2 mitigation but also very likely on other prioritized governmental affairs. China is also applying the same strategy in governing CO_2 mitigation. Other countries may also find this alternative governance model helpful in contributing solutions to their major public problems.

References

Barrett, S. & Graddy, K. 2000. Freedom, growth, and the environment. *Environment and Development Economics*, 5, 433–456.

Beeson, M. 2010. The coming of environmental authoritarianism. *Environmental Politics*, 19, 276–294.

BP. 2019. *Statistical review of world energy* [Online]. Available: https://www.bp.com/content/dam/bp/business-sites/en/global/corporate/pdfs/energy-economics/statistical-review/bp-stats-review-2019-full-report.pdf.

Buitenzorgy, M. & Mol, A. P. J. 2011. Does democracy lead to a better environment? Deforestation and the democratic transition peak. *Environmental & Resource Economics*, 48, 59–70.

Congleton, R. D. 1992. Political-institutions and pollution-control. *Review of Economics and Statistics*, 74, 412–421.

Crippa, M., Guizzardi, D., Muntean, M., Schaaf, E., Dentener, F., Van Aardenne, J. A., Monni, S., Doering, U., Olivier, J. G. J., Pagliari, V. & Janssens-Maenhout, G. 2018. Gridded emissions of air pollutants for the period 1970–2012 within Edgar v4.3.2. *Earth System Science Data*, 10, 1987–2013.

Downey, L. & Strife, S. 2010. Inequality, democracy, and the environment. *Organization & Environment*, 23, 155–188.

EIA. 2019. *International energy statistics*. Washington, DC: U.S. Energy Information Administration.

Fioletov, V., McLinden, C., Krotkov, N., Li, C., Leonard, P., Joiner, J. & Carn, S. 2019. *Multi-satellite air quality sulfur dioxide (SO2) database long-term L4 global V1*. Goddard Earth Science Data and Information Services Center (GES DISC) [Online]. Available: https://disc.gsfc.nasa.gov/datasets/MSAQSO2L4_1/summary.

Fridley, D. & Lu, H. 2016. *China energy databook version 9.0*. Berkeley, CA: Lawrence Berkeley National Laboratory.

Grossman, G. M. & Krueger, A. B. 1995. Economic-growth and the environment. *Quarterly Journal of Economics*, 110, 353–377.

Hagene, T. 2010. Everyday political practices, democracy and the environment in a native village in Mexico city. *Political Geography*, 29, 209–219.

IMF. 2019. *World economic outlook database October 2019* [Online]. Available: https://www.imf.org/en/Publications/SPROLLS/world-economic-outlook-databases#sort=%40imfdate%20descending.

Institute for Health Metrics and Evaluation. 2018. *Global burden of disease (GBD) 2017 study* [Online]. Available: https://www.thelancet.com/journals/lancet/article/PIIS0140-6736(18)32279-7/fulltext.

Kaufmann, D. & Kraay, A. 2019. *The worldwide governance indicators 2019 update: Aggregate governance indicators 1996–2018* [Online]. Available: https://papers.ssrn.com/sol3/papers.cfm?abstract_id=1682130.

Li, C., McLinden, C., Fioletov, V., Krotkov, N., Carn, S., Joiner, J., Streets, D., He, H., Ren, X. R., Li, Z. Q. & Dickerson, R. R. 2017. India is overtaking China as the world's largest emitter of anthropogenic sulfur dioxide. *Scientific Reports*, 7.

Li, Q. & Reuveny, R. 2006. Democracy and environmental degradation. *International Studies Quarterly*, 50, 935–956.

Lu, Z., Zhang, Q. & Streets, D. G. 2011. Sulfur dioxide and primary carbonaceous aerosol emissions in China and India, 1996–2010. *Atmospheric Chemistry and Physics*, 11, 9839–9864.

Marshall, M. G., Gurr, T. R. & Jaggers, K. 2019. *Polity IV project: Political regime characteristics and transitions, 1800–2018*. Center for Systemic Peace [Online]. Available: http://www.columbia.edu/acis/eds/data_search/1080.html.

Midlarsky, M. I. 1998. Democracy and the environment: An empirical assessment. *Journal of Peace Research*, 35, 341–361.

National Bureau of Statistics. 2019. *China statistical yearbook*. Beijing, China: China Statistics Press.

National People's Congress. 2006. *The outline of the national 11th five-year plan on economic and social development*. Beijing, China: The 4th Conference of the 10th National People's Congress.

National Statistics Bureau & Ministry of Ecology and Environment. 2019. *China statistical yearbook on environment 2018*. Beijing, China: China Statistics Press.

Neumayer, E. 2002. Do democracies exhibit stronger international environmental commitment? A cross-country analysis. *Journal of Peace Research*, 39, 139–164.

Njeru, J. 2010. 'Defying' democratization and environmental protection in Kenya: The case of Karura forest reserve in Nairobi. *Political Geography*, 29, 333–342.

Payne, R. A. 1995. Freedom and the environment. *Journal of Democracy*, 6, 41–55.

Pellegrini, L. & Gerlagh, R. 2006. Corruption, democracy, and environmental policy – an empirical contribution to the debate. *The Journal of Environment & Development*, 15, 332–354.

Stern, D. I. & Common, M. S. 2001. Is there an environmental Kuznets curve for sulfur? *Journal of Environmental Economics and Management*, 41, 162–178.

Sundberg, J. 2003. Conservation and democratization: Constituting citizenship in the Maya biosphere reserve, Guatemala. *Political Geography*, 22, 715–740.

Torras, M. & Boyce, J. K. 1998. Income, inequality, and pollution: A reassessment of the environmental Kuznets curve. *Ecological Economics*, 25, 147–160.

U.S. EPA. 2019. *Air pollutant emissions trends data: Criteria pollutants national tier 1 for 1970–2018*. Washington, DC: U.S. EPA.

Walker, P. A. 1999. Democracy and environment: Congruencies and contradictions in Southern Africa. *Political Geography*, 18, 257–284.

Wendling, Z. A., Emerson, J. W., Esty, D. C., Levy, M. A., De Sherbinin, A. et al. 2018. *2018 environmental performance index*. New Haven, CT: Yale Center for Environmental Law & Policy.

Winslow, M. 2005. Is democracy good for the environment? *Journal of Environmental Planning and Management*, 48, 771–783.

Zheng, B., Tong, D., Li, M., Liu, F., Hong, C. P., Geng, G. N., Li, H. Y., Li, X., Peng, L. Q., Qi, J., Yan, L., Zhang, Y. X., Zhao, H. Y., Zheng, Y. X., He, K. B. & Zhang, Q. 2018. Trends in China's anthropogenic emissions since 2010 as the consequence of clean air actions. *Atmospheric Chemistry and Physics*, 18, 14095–14111.

2 Political will

1 Centralized political will

Which governmental affairs can become national priorities and their relative rankings are highly centralized in the Chinese context without democracy. In contrast to the path argued by Payne (1995), in which a democracy develops its political will regarding the environment, China has taken a different route. The state is far more dominant in China than it is in a democracy. Even nongovernmental organizations (NGOs) in China actively seek alliances with the government (Hsu, 2010). The lack of free elections also reduces the need for the government to directly respond to the public's demands.

The Chinese Communist Party holds tremendous authority in deciding, for example, how important environmental protection is among all governmental affairs. The party is closely intertwined with the Chinese government, but they are also very different. The party makes key decisions while the government takes almost all implementation tasks. Although the party has about 90 million members and is organized into multiple levels, the authority is very much centralized upward and eventually into the Central Committee. The 19th cohort was inaugurated in October 2017 after the corresponding National Party's Congress. It has 204 members, and their tenure will last for five years, until 2022 when the next National Party's Congress convenes to form another Central Committee. It further forms the Political Bureau, currently with 25 members, and then, most crucially, the 7-member Standing Committee as China's top leadership. Many of these members, but not all, also hold positions in the Chinese government. Two are most important. The secretary general, currently Xi Jinping, is at the center and generally assumes the position of president in the Chinese government. The prime minister, currently Li Keqiang, leads the Chinese administration. This hierarchy ensures China's high degree of centralization in making most important decisions. The Chinese government and, specifically, environmental administration are mainly focused on environmental policy making and implementation. On those prioritized governmental affairs that decisions have been made by the top leadership of the Party, the government is in charge of implementation.

In the past seven decades after the establishment of the People's Republic of China, each top leadership of the Chinese Communist Party has left a phrase in the party's Constitution, with their ideologies written as the party's "guiding compass," which not only guides their own leadership's rule but also summarizes a legacy. The line has become longer over time to include "Mao Zedong Thoughts," "Deng Xiaoping Theory," "Three Representativeness" (headed by President Jiang Zemin), "Scientific View of Development" (headed by President Hu Jintao) and "Socialistic Thoughts with Chinese Characteristics in the Xi Jinping Era" (Chinese Communist Party, 2017).

This chapter mainly focuses on how the political will for environmental protection has evolved since the 15th Central Committee was formed in 1998. The period transcended three top leaderships of the party, including President Jiang Zemin and Prime Minister Zhu Rongji (1998–2002), President Hu Jintao and Prime Minister Wen Jiabao (2003–2012) and President Xi Jinping and Primer Minister Li Keqiang (2013–2022). Chapter 3 examines the environmental governance of the Chinese government for implementing the political will.

2 Economy, jobs and the environment (1998–2002)

The period was under the 15th Central Committee and the leadership of President Jiang Zemin and Prime Minister Zhu Rongji. Although China's environmental pollution had already reached high levels, more urgent issues were present to suppress forceful political will for environmental protection. Difficult economic conditions slowed down energy consumption to witness a decline of sulfur dioxide (SO_2) emissions in the 9th Five-Year Plan (1996–2000; Figure 1.8).

China's economy was still at the early stage of industrialization, while the Asian financial crisis of 1997 hit China badly. In comparison with the previous years (1992–1997), the average annual gross domestic product (GDP) growth rate declined significantly from 11.8% to 8.3% (Figure 2.1). GDP per capita was still at low levels, US$3,185 (purchasing power parity [PPP] in 2011 US$) in 1998 and US$4,276 in 2002, or 7.4% and 9.3% of the U.S. levels, respectively (Figure 1.6). Job creation was more important than GDP growth. As will be introduced in Chapter 3, the following year, 1998, witnessed China's several far-reaching fundamental reforms with a key focus on state-owned enterprises and a better-defined boundary between the state and the market. Many of these state-owned enterprises were substantially overstaffed and loss-making and operated more like governmental agencies and less like market-oriented entities. The Chinese financial sector and, specifically, the state-owned banks had extremely high levels of bad debts. This period witnessed large-scale privatization and the bankruptcy of small and medium-sized state-owned enterprises, mainly in the secondary sector. As a result, the secondary sector shed 8.7 million jobs from 1998 to 2002 to reflect the massive reform's side effects (Figure 2.1). Overall, 3.3 million jobs were annually added to the secondary and tertiary sectors. With many more people entering than leaving the workforce as indicated in the rapidly enlarging age group of 15- to 64-year-olds (Figure 2.2), many of the unemployed should have returned to rural regions as the primary sector added 18.0 million jobs over the five years (Figure 2.1). China's job and demographic structures were still dominated by the

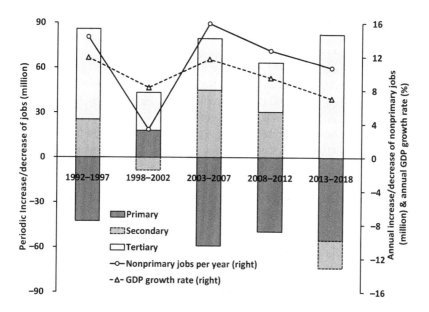

Figure 2.1 Sectoral employment changes and GDP growth rates across China's administrations
Source: National Bureau of Statistics (2019).

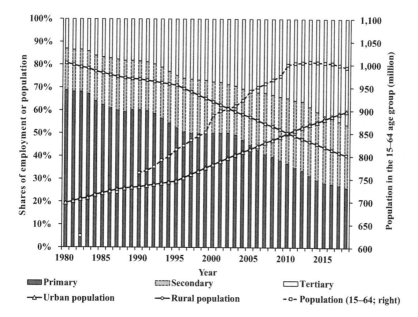

Figure 2.2 Employment and population structures in China
Source: National Bureau of Statistics (2019).

primary sector and rural regions. The primary sector's share of total jobs hovered stably between 49.8% and 50.1%, while the share of the rural population declined from 68.1% in 1997 to 60.9% in 2002 (Figure 2.2). As a result, most Chinese were less exposed to seriously polluted urban air pollution because they were not breathing urban air.

Generally speaking, over the period from 1998 to 2002, environmental protection was ranked high neither in governmental affairs nor by society. In the aftermath of the Asian financial crisis, economic downturn and unemployment were more imminent and highly politicized problems to occupy the top leadership's mind. This top leadership's guiding ideology, "three representativeness," was mainly engaged in expanding the party's base from the conventional working class to other categories of the society. Environmental protection did not occupy any important role in this ideology, while slower industrial development also reduced the deterioration rate of environmental pollution.

3 SARS and the prioritization of environmental protection (2003–2012)

Over the ten years (two terms with the 16th and 17th Central Committee) between 2003 and 2012, when President Hu Jintao and Prime Minister Wen Jiabao were in power, China added 75.6 million new jobs in the secondary sector and 67.3 million in the tertiary sector, while the primary sector had a decrease of 108.7 million jobs (Figure 2.1). To keep pace with the growing working-age population, the annual increase of nonprimary jobs was 14.3 million, much faster than the 3.3 million new jobs annually between 1998 and 2002 (Figure 2.1). The primary sector still accounted for 50.0% of China's overall employment in 2002 and remained the largest among the three sectors in 2007 at 40.8%. China's entry into the World Trade Organization in 2001 and multiple major economic reforms led to unprecedented growth in the economy, energy consumption and pollution. The global financial crisis of 2008 did exert great and negative impacts on China's economy to slow it down. Comparing the two Hu-Wen administrations (2003–2007 and 2008–2012), the annual economic growth rate came down from 11.7% to 9.4%, and the annual increase of nonprimary jobs was from 15.9 million to 12.7 million.

Environmental protection started to emerge as a nationally prioritized governmental affair. The 11th Five-Year Plan (2006–2010) was completely formulated and implemented under this top leadership of the party. It not only included the 10% mitigation goals of SO_2 and chemical oxygen demand but actually achieved them (National People's Congress, 2011), defying challenges from the rapid growth of economy and energy consumption and reversing the humiliating failures in the 10th Five-Year Plan (Figure 1.8). Deeper mitigation of SO_2 emissions followed in later years, while the turning point of environmental protection happened within this period (Figure 1.8).

Society might not have been ready to put the environment as a high priority with strong cleanup determination. For example, despite the dire situation of air pollution, a survey in 2010 by Gallup, a U.S. research-based consulting company, found that only 26% of the Chinese were dissatisfied, and 73% were satisfied, with the air quality (English, 2010). The potentially insufficient support from society for pollution mitigation, if China were a democracy, might not have generated strong political will.

The much stronger political will for environmental protection reflected more the intention of the top leadership of the party. As is examined in detail in Chapter 4, the direct involvement of the top leadership was crucial in enacting the environmental goals in the 11th Five-Year Plan after the failures in the 10th Five-Year Plan. The 16th Central Committee was formed in November 2002 at the 16th National Party's Congress. The Standing Committee of the Political Bureaus was headed by Secretary General Hu Jintao and included Wen Jiabao. In March 2003, at the 10th National People's Congress, they assumed the positions of president and prime minister, respectively, in the Chinese central government. In the transitional period between these two key conferences, they had only party leadership roles but officially not those later government positions.

SARS (severe acute respiratory syndrome), a new infectious disease, emerged almost exactly over this transitional period in November 2002 and became increasingly damaging over the winter (WHO, 2003). The timing of the devastating pandemic coincided well with the top leadership's search for a new ideology to distinguish themselves from their predecessors. This public health crisis taught a painful lesson to the Chinese leadership that public goods should be prioritized together with economic development. The overemphasis of the latter may actually backfire to result in slow economic growth as the Chinese economy was significantly damaged, especially in the second quarter of 2003, by the impacts of the SARS pandemic (Rawski, 2005; Hai et al., 2004; Xu et al., 2009). After the pandemic was over and society returned to normal, a new ideology was gradually formed, titled "Science View of Development," to emphasize development from multiple aspects to achieve a "harmonious society." Environmental protection is a natural extension from public health and became one pivotal component in this new development direction.

The authorities of the top leadership and this new ideology were hardly distinguishable. From this perspective, whether China could achieve serious mitigation of environmental pollution and reverse the deterioration trend became more politicized. This significantly increased the political will of the top leadership to start taking environmental protection into the inner core of key governmental affairs. In other words, the political will resulted from a more top-down rather than bottom-up approach, although the pressure from society grew over the years.

Key international events also played a role in shaping China's environmental protection. One of the most important events over the Hu–Wen administrations was the 29th Summer Olympic Games in August 2008. To ensure good air quality over Beijing, China shut down many polluting factories across several neighboring provinces around Beijing. Environmental information was increasingly available over this period. The Internet played a key role in distributing information. The U.S. Embassy in Beijing started monitoring fine particulate matter ($PM_{2.5}$) levels in 2008. Environmental NGOs, notably the IPE (Institute of

Public & Environmental Affairs) that was established in 2006, started systematically collecting, publicizing and distributing environmental information to the public.

4 The sustainability of environmental political will (2013–present)

President Xi Jinping and Prime Minister Li Keqiang assumed their top leadership roles of the Chinese Communist Party in November 2012 at the 18th National Party's Congress and then of the central government in March 2013 at the 12th National People's Congress. As usual, the change of leadership did raise questions about whether environmental protection could be further strengthened or weakened in relation to new economic conditions and new leaders' ideas. The Chinese economy entered a "new normal," or a stabilized but lower level after 2013. The annual GDP growth rate from 2013 to 2018 was 7.0%, even lower than the level during the aftermath of the Asian financial crisis. Nevertheless, the economy had already reached a wealthier status before the new leadership came into power and the progress since 2013 has also been decent. In 2002, China's GDP per capita was US\$4,276 (PPP in 2011 US\$), and it increased to US\$11,049 in 2012 and US\$16,098 in 2018 (IMF, 2019). The ratios between China and the United States were 9.3%, 21.8% and 28.8%, respectively.

With the working-age population stabilized at about 1 billion people (Figure 2.2), job creation was still at a healthy pace with 10.7 million new nonprimary jobs added annually. Over the six years, in total, the tertiary sector added 82.5 million new jobs, while the secondary and primary sectors had 18.5 million and 55.2 million fewer jobs (Figure 2.1). In contrast to the economic downturn between 1998 and 2002, Chinese labor did not return to rural regions. The tertiary sector accelerated significantly to account for 46.3% of all employment in 2018, up from 36.1% in 2012 (Figure 2.2). The primary sector accounted for 31.4% of all jobs in 2013 and further declined to only 26.1% in 2018 (Figure 2.2). Furthermore, China has been urbanizing fast to have 53.7% of people in urban regions in 2013. In 2018, the urbanization rate further increased to 59.6% (Figure 2.1). In other words, China's employment and demographic structures have been much more urbanized, which also brought more people under the impacts of more polluted urban air.

Rapid economic development and escalating living standards have been key foundations for the Chinese people to maintain support to the Chinese Communist Party's holding of power. The Chinese middle class has expanded rapidly in the past decades to indicate that this demand was to a great extent satisfied. Given the higher income and more intimate exposure of an average Chinese to urban air pollution, society started to place environmental quality at a significantly higher priority than before. The balance between environmental protection and economic growth has thus been shifting gradually toward the former's end. In the leadership transitional period in January 2013, North China suffered from severe smog with $PM_{2.5}$ concentration levels reaching hazardous levels (Wang et al., 2014).

Although Hebei Province had worse air quality, it was Beijing, as China's capital, that attracted most international and domestic attention. Air pollution mitigation started to be widely recognized as one crucial demand by society. People are increasingly willing to sacrifice economic opportunities for a better environment. Environmental protection and especially urban air quality have been significantly politicized, now by society, and implicitly linked with the legitimacy of the Chinese Communist Party as the ruling political party.

In addition, environmental protection also became a more and more visible business to create jobs and economic outputs. The initial efforts in the Hu–Wen administrations started to bear fruits. China's environmental and renewable energy industries are competitive not only domestically but also internationally (Xu, 2013; Zhu et al., 2019). They have grown into another pollical force to push for China's continuous environmental cleanup. For example, China now has the world's largest solar, wind and electric vehicle industries. They play increasingly counterbalancing roles against those who are concerned about the negative impacts of environmental protection on their businesses.

In the formation of this top leadership's governing ideology, the party was also keen to significantly elevate the priority of environmental protection. The 18th National Party's Congress in 2012 emphasized ecological civilization, while the 19th National Party's Congress in 2017 listed "harmony of people and nature" as one of the 14 basic things to insist on, which primarily features ecological civilization and the "two mountains" theory. Previously, in the relationship between economic development and environmental protection, the statement was that we want not only "gold and silver mountain" but also "clear water and green mountain." In other words, these two were placed as trade-offs to each other. The new statement of "two mountains" became that "clear water and green mountain" are "gold and silver mountain." The pursuit of environmental quality became equivalent to economic development. Environmental protection does offer opportunities to satisfy the demands for both economic development and a better environment, for example, when new industries emerge for pollution mitigation or resource conservation. Environmental policies have also been playing an active role in encouraging innovation and economic transformation, as elaborated in greater detail in Chapter 7.

Overall, in this period, both the top leadership of the party and society came together with a common and prioritized stake in a cleaner environment. Environmental protection is increasingly politicized to form an unprecedented political will for pollution mitigation. The top leadership should meet the growing demand of the society for not just economic growth but also environmental cleanup. Because "ecological civilization" is a key component in the top leadership's "Socialistic Thoughts with Chinese Characteristics in the Xi Jinping Era," significant improvement of environmental quality also became crucial for the establishment of this new governing ideology. New economic opportunities and environmental industries have been serving as an increasingly visible force to counterbalance the negative economic impacts of environmental protection. The rapid growth of income has also transformed society's preference between economic development and environmental quality. They are crucial forces to make the political will sustainable, even when top leadership changes again in the future.

References

Chinese Communist Party. 2017. *The party's constitution (Revised by the 19th national party's congress)* [Online]. Available: https://www.researchgate.net/publication/322421053_The_19th_Congress_of_the_Communist_Party_of_China_and_Its_Aftermath.

English, C. 2010. *More than 1 billion worldwide critical of air quality*. Washington, DC: Gallup.

Hai, W., Zhao, Z., Wang, J. & Hou, Z. G. 2004. The short-term impact of SARS on the Chinese economy. *Asian Economic Papers*, 3, 57–61.

Hsu, C. 2010. Beyond civil society: An organizational perspective on state – NGO relations in the people's Republic of China. *Journal of Civil Society*, 6, 259–277.

IMF. 2019. *World economic outlook databases October 2019* [Online]. Available: https://www.imf.org/en/Publications/SPROLLs/world-economic-outlook-databases#sort=%40imfdate%20descending.

National Bureau of Statistics. 2019. *China statistical yearbook*. Beijing, China: China Statistics Press.

National People's Congress. 2011. *The outline of the national 12th five-year plan on economic and social development*. Beijing, China: The 4th Conference of the 10th National People's Congress.

Payne, R. A. 1995. Freedom and the environment. *Journal of Democracy*, 6, 41–55.

Rawski, T. G. 2005. SARS and China's economy. *In:* Kleinman, A. & Watson, J. L. (eds.) *SARS in China: Prelude to pandemic?* Stanford: Stanford University Press.

Wang, Y. S., Yao, L., Wang, L. L., Liu, Z. R., Ji, D. S., Tang, G. Q., Zhang, J. K., Sun, Y., Hu, B. & Xin, J. Y. 2014. Mechanism for the formation of the January 2013 heavy haze pollution episode over central and eastern China. *Science China-Earth Sciences*, 57, 14–25.

WHO. 2003. *Update 95 – SARS: Chronology of a serial killer* [Online]. Available: https://www.who.int/csr/don/2003_07_04/en/.

Xu, Y. 2013. Comparative advantage strategy for rapid pollution mitigation in China. *Environmental Science & Technology*, 47, 9596–9603.

Xu, Y., Williams, R. H. & Socolow, R. H. 2009. China's rapid deployment of SO2 scrubbers. *Energy & Environmental Science*, 2, 459–465.

Zhu, L., Xu, Y. & Pan, Y. J. 2019. Enabled comparative advantage strategy in China's solar PV development. *Energy Policy*, 133.

3 Environmental governance

1 Evolution of environmental administration

Environmental protection in China could be traced back to the United Nations Conference on the Human Environment in June 1972 in Stockholm, Sweden. In the turmoil of the Cultural Revolution (1966–1976) and after the United Nations voted in 1971 that the People's Republic of China is the sole representative of China, China sent an official delegation to this conference. In August 1973, the First National Conference on Environmental Protection was held to mark that environmental protection had formally been recognized as a governmental affair. However, in the early stage of the Cultural Revolution, the leaders and organizations of the Chinese Communist Party and the Chinese government at various levels were generally toppled by Red Guards (*hong wei bin*) and Rebels (*zao fan pai*). Although the Chinese government was rebuilt at a later stage, the primary focus was not on economic or social affairs but on class struggle. As a result, China did not demonstrate a significant conflict between economic development and environmental protection because neither mattered.

When the Cultural Revolution ended in 1976, after a short transitional period, China entered the new era of Reform and Open-up in December 1978. Economic development quickly gained prominence in governmental affairs, while class struggle and other political affairs wound down. Soon afterward, the impacts of economic development on environmental quality started to emerge. As a public affair that requires governmental intervention, environmental protection was announced as one Basic National Policy in the Second National Conference on Environmental Protection from 31 December 1983 to 7 January 1984. Since then, dedicated governmental entities have been established in the Chinese government to regulate and implement environmental protection. The agency in the Chinese central government that oversees environmental protection has evolved over the years in terms of organization, power and jurisdiction. The authority of environmental protection has been increasingly strengthened in the past four decades. In 1984, the State Environmental Protection Agency was established under the then Ministry of Construction. In 1988, it was pulled out to be directly led by the State Council, thus with an elevated status and authority at the vice-ministry level. Environmental protection then became not just an issue for one single ministry but also one key state affair that was widely relevant and one level closer to the center of the governmental authority.

China's key reforms in the past four decades have one crucial central theme for adjusting the relationship between the state and the market. There were essentially no real markets in the Cultural Revolution because markets were deemed as too capitalistic. Prices did not reflect any balance between demand and supply but were decided directly by the government. Purchases should be accompanied by permits, not just money. Despite fluctuations, the overall trend in the past decades was the reemergence, creation and maturity of various markets, as well as the refocusing of the state from everything to strategic and public affairs. With the government giving up its original authority, prices have become much better indicators of supply and demand balances. The production and consumption are increasingly guided by market signals and little by orders from central planners.

In the 1998 reform of the State Council that featured a better-clarified demarcation between the state and the market, 14 ministries that mainly took direct charge of the economic sectors were abolished, and 4 new ministries were formed. The government then became more focused on public affairs and much less on direct management of businesses. In this reform, the then State Environmental Protection Agency was promoted to the ministerial level and renamed the State Environmental Protection Administration (SEPA). Other significant reforms in the same period marked the reorganization of large state-owned enterprises, the privatization of small ones and the widened space for private businesses.

Although environmental protection gained increasingly higher statuses in the previously mentioned reforms, it was still kept away from the core of the Chinese central government, in which the State Council is in charge of the country's routine administration. According to China's Constitution, the State Council comprises the following members: prime minister and deputies, state councilors, ministers, directors of commissions and the auditor general. Although the SEPA had been elevated to the ministerial level after the 1998 reform, it was not a ministry, and thus, its director was not a constitutional member of the State Council. He or she could be present in the meeting only by invitation, with much constrained authority on other ministries' affairs even if they may be closely relevant to environmental protection. The 2008 reform became crucial when the SEPA was reorganized as the Ministry of Environmental Protection and thus became a formal comprising ministry of the State Council. This reform indicated that environmental protection was recognized as one of the key governmental affairs. The enhanced authority also gave the new ministry and its counterparts in local governments more forceful power in enacting and implementing environmental policies.

In 2018, a new round of major reforms further concentrated environmental authorities that scattered in several ministries into the newly formed Ministry of Ecology and Environment (MEE; State Council, 2018). Climate change was notably transferred out of the National Development and Reform Commission to fall under the MEE's jurisdiction. The MEE now combines the original functions of (1) Ministry of Environmental Protection, (2) climate change and mitigation under the National Development and Reform Commissions, (3) groundwater pollution under the Ministry of Land and Resources, (4) water environment management under the Ministry of Water Resources, (5) agricultural pollution under the Ministry of Agriculture, (6) ocean environment under the State Oceanic Administration and (7) south–north water diversion project's environmental protection under its office. This reform further strengthened the authority of environmental protection. The significantly wider duties are expected to create better synergies among their regulations and solutions.

2 Chain of command for environmental protection

Environmental protection administration in China has four major levels, being central, provincial, municipality and county. The latter three levels are generally categorized as local governments, although provincial governments are often not directly involved in local administration. Local governments take primary responsibilities for implementing environmental policies and achieving environmental protection. The sequential reforms at the central government were followed by corresponding reforms in local governments that generally resemble the structures of the central government, despite differences contingent on local contexts.

Although the MEE and its predecessors had a clear chain of command under the State Council of the central government, it is not straightforward whether local environmental protection bureaus (EPBs) should be led by corresponding local governments or environmental protection agencies at a higher governmental level for achieving more effective environmental administration. On one hand, environmental protection is far beyond the authority of the EPBs to involve industrial policy, urban planning and other policies. Environmental enforcement heavily relies on other agencies and budget allocation from local governments. Accordingly, it is reasonable to have local governments as the major office-bearers. On the other hand, local governments may create barriers to environmental protection due to the possible conflicts between economic growth and environmental protection. If local EPBs could be vertically controlled, they may better serve the purpose of environmental protection as local economic growth is not the central consideration of upper-level EPBs.

China's administrative reform in the past four decades has one key trend: more and more remaining governmental authorities are being decentralized from the central government to local governments, especially regarding the regulation of economic activities and the provision of social public goods such as health care, education, housing and urban/rural infrastructure and community services. In the environmental administrative system, the chain of command for local EPBs reflected such a decentralization trend to recognize that environmental protection is generally a localized governmental affair. In 1999, the Department of Organization of the Chinese Communist Party reformed the institutional arrangements and specified that the leaders of local EPBs should be jointly appointed by primarily local governments and, to a lesser extent, upper-level EPBs (Department of Organization of the Central Committee of the Communist Party of China, 1999). The "double administration" arrangement aimed for a balance between the vertical – or "*tiao*" based on the function of environmental administration – and horizontal – or "*kuai*" based on the location of environmental protection. The 1999 reform was accordingly mainly horizontally oriented with decentralization. EPBs were under local governments with their directors and budgets controlled by their corresponding local governments. They were also advised by EPBs in the immediate upper-level governments.

The general decentralization trajectory in the past decades also engaged another argument for recentralization. In the era of Reform and Open-up, local governments often have to face the conflicts between environmental protection and economic development. In evaluating the performance of local leaders, economic indicators tended to occupy much heavier weights than environmental protection, especially in the early years. Accordingly, for the sake of the local economy, the environment has often been sacrificed. Together with the rising status of environmental protection in the central government as described earlier, environmental protection started to climb higher on the priority list. The MEE as well as its predecessors and local counterparts are less bound by such evaluation because economic development is not their direct job duty, but environmental protection is their primary responsibility. In 2016, another major and more centralization-oriented reform was initiated with several provinces for pilot implementation (The General Office of the CPC Central Committee and The General Office of the State Council, 2016). The authority of appointing local EPB leaders and their budgets were shifted more toward upper-level EPBs. Environmental monitoring and inspection agencies were more directly controlled vertically.

3 Division of labor for policy making and implementation

Environmental agencies in China's central and local governments have distinct functional focuses. The central government is mainly in charge of policy making. It also supervises local governments, primarily provincial governments, for implementing environmental protection. Provincial governments heavily focus on policy making within their individual provinces. They also adapt policies from the central government to their own situations and supervise mainly municipality governments. The municipality level has a further diminished capacity in policy making and a much heavier focus on policy implementation, while the tasks of county governments fall almost exclusively on the implementation of policies from the upper levels within localized contexts. Implementation is primarily the responsibility of municipality and county governments. They can also make decisions that are applied within their specific jurisdictions, mainly on how to implement policies with greater efficiency and effectiveness.

The clear division of labor among the four levels of governments is reflected in their composition of environmental protection personnel. Their personnel compositions are accordingly different among the four categories: administration, inspection, monitoring and others. "Administration" mainly refers to the MEE in the central government as well as corresponding bureaus at the three levels of local governments. "Inspection" personnel are those who work in Inspection Bureaus, while "monitoring" personnel are based in Monitoring Stations. "Others" are the remaining personnel, such as those in the Academy of Environmental Sciences and Academy of Environmental Planning at the four levels. They provide research and expertise to support environmental policy and decision making. Between 2004 and 2015, using available data, the compositions at the four governmental levels were largely stable (Figure 3.1). The only significant exception is the share of "inspection" at the central level, which experienced a dramatic increase in 2009 (Figure 3.1).

The environmental authority in the central government is not organized for shouldering implementation tasks but primarily for making policies and supervising local governments (SCOPSR, 2018). At the central level, "others" is the largest category. It accounted for 64.1% of all 3,023 environmental protection personnel in 2015, while the share was over 80% before 2009 (Figure 3.1). Their dominant share indicates that environmental policy making in China requires and has been receiving significant intellectual support. "Administration" hosted only 362 personnel in 2015, and its share remained stable at about 12% over the period between 2004 and 2015 based on available data. After the 2018 reform and the reorganization, the new MEE was allowed to have 478 personnel, the addition for accommodating expanded functions (SCOPSR,

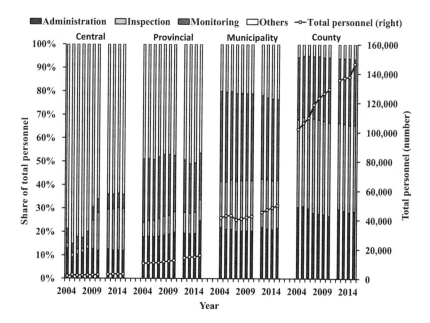

Figure 3.1 Environmental protection personnel at four governmental levels in China (for 2004–2015 using available data)

Source: Ministry of Environmental Protection (2002–2016).

2018). Partly as a result of the establishment of six Regional Supervision Centers, "inspection" had a major shift with its personnel jumping from 41 in 2008 to 294 in 2009 and further to 542 in 2015. The 2018 reform further formalized and upgraded them into Regional Supervision Bureaus, with a total personnel capacity of 240 officers (SCOPSR, 2018). "Monitoring" had about 6% of all personnel throughout the years. The inspection and monitoring personnel provide crucial data support for supervising the environmental protection performance of local governments.

China's provincial environmental authorities are also structured to have a heavy focus on policy making and supervision and less a focus on direct policy implementation. At the provincial level, "others" remains the largest to have 46.4% of all provincial environmental protection personnel in 2015 (Figure 3.1). It is the largest category to reflect the desired functions in policy making. "Monitoring" occupied 19.9%, which was a decline from 26.9% in 2004 (Figure 3.1). The share of "inspection" increased from 6.2% in 2004 to 9.0% in 2015, but the increase was much less significant in comparison with that at the central level (Figure 3.1). Between monitoring and inspection, the central government now puts more emphasis on inspection while provincial governments have a heavier focus on monitoring.

The municipality and county levels are structured with much lower capacities for policy making and primarily for policy implementation. At the municipality level, "monitoring" is the largest category, with 34.5% of all its environmental protection personnel in 2015 (Figure 3.1). "Inspection," "administration" and "others" each took about one fifth of the personnel. Their primary tasks are, accordingly, sharply different from the central and provincial levels, with a heavy focus on actually implementing policies, although they also build decent knowledge support for initiating policy innovations. The county level is almost exclusively for implementation, with "others' accounting for only 6.4% of environmental protection personnel in 2015. "Inspection" became the largest functional group with 37.0% of personnel, while "monitoring" and "administration" had 28.0% and 28.6%, respectively (Figure 3.1).

The differentiated functions of environmental authorities at the four levels indicate that China's environmental protection requires their close cooperation. From the MEE in the central government to environmental protection bureaus at the county level, policy making is more concentrated at the top while implementation is mainly at the lower levels. However, their cooperation should not be taken for granted, even though China has a conventional image of top-down administration. As examined in later sections, local governments and their leaders have their own self-interests. If environmental policy implementation is against such interests, the implementation will not be expected to be effective. As expected from China's weak rule of law, regardless of how stringent environmental policies are, their weak implementation was one of the primary reasons that led to China's environmental crises. Without forceful enforcement efforts of local governments and widespread compliance of polluting sources, environmental cleanup cannot be realized.

4 Decentralized policy making

From social, economic, industrial and environmental perspectives, China has been evolving at an astonishing speed in the past four decades. Laws, policies and regulations should continuously adapt to the rapidly changing situations. As indicated in the World Bank's governance indicators, the rule of law in China has not been well established (Kaufmann and Kraay, 2019). Laws and courts have not been playing important roles in daily environmental protection. Instead, policies and regulations are much more closely relevant.

Laws in China are enacted by the National People's Congress. They tend to take many years to formulate, enact or amend. For example, the Law of Environmental Protection is the basic law to regulate China's environmental protection. It was first enacted in 1989, and then it took 25 years to get amended in 2014. However, China's environmental conditions and pollution had dramatically changed during the 25 years, which should have indicated that the older version was seriously outdated. In addition, a variety of specific laws are enacted to regulate individual categories of the environment. For example, the Law of Atmospheric Pollution Prevention and Control was enacted in 1987, and two amendments have been done since then, in 2000 and 2015 (two other minor corrections were done in 1995 and 2018; National People's Congress, 2018). The Law of Water Pollution Prevention and Control was enacted in 1984. Only one amendment has been done in 2008, while two minor corrections were made in 1996 and 2017 (National People's Congress, 2017). Accordingly, many environmental policies in China do not have clear corresponding items in environmental laws.

The slow motion of laws' enactment and amendment may make them at a great distance from the rapidly evolving pollution conditions. This could also partly explain why many of China's policies were applied before their legal foundations were established. For example, the eco-compensation policy got its legal backing only in 2014 in the newly amended Environmental Protection Law, but by then, it had already been experimented with and applied widely (Wang et al., 2016). Furthermore, courts do not play significant roles in environmental enforcement and compliance. The laws are often written to mainly state principles without enough details for direct implementation. The situations reflect China's situation of weak rule of law. The Chinese central government does not file lawsuits against local governments for not implementing laws and its policies.

In addition, China's environmental laws are often intentionally vague in order to allow more flexibility for the administration, while environmental policies contain more implementable details. Compared with the U.S. Clean Air Act Amendments (CAAA, 1990), China's goal and initial plan were much less detailed. The CAAA clearly developed a cap-and-trade system with detailed rules and schedules (The U.S. Congress, 1990). Such details were absent in China's plans. China's laws are often drafted by a ministry, not the National People's Congress. For example, a key task of the MEE is to draft laws and regulations on environmental protection (SCOPSR, 2018). A vague law can provide a legal foundation but not constrain the enactment of policies. For example, China's Law of Atmospheric Pollution Prevention and Control entitles the environmental authority to enact ambient air quality standards and effluent emission standards without further clarification on when and how (National People's Congress, 2000). The State Council gets the legal power to collect effluent emission charges and the freedom to enact any relevant regulation (National People's Congress, 2000).

Other than the National People's Congress, the State Council can enact Regulations. Various ministries, as well as their internal departments, frequently churn out policies, standards, projects and other incentives/commands that are relevant to environmental protection. (For simplicity, they are referred to as environmental policies in the following discussion.) Local governments and their environmental authorities also hold the right to enact their own environmental policies or to adapt those from the central government into their corresponding jurisdictions and contexts. As shown in Figure 3.1, local governments, especially at the provincial and (to a lesser extent) municipality levels, do have decent capacities for making policies. All these environmental policies could have very different scopes, stringency, instruments, targets and intellectual support. In comparison to laws, environmental policies are much more flexible. Its enactment takes much less time and faces much lower hurdles. The entire process is also much less centralized with numerous governmental bodies at ministerial and local levels who can independently enact environmental policies. China's weak rule of law indicates that these policies are rarely challenged in courts or through other channels by affected interest groups, although their legal foundation might be porous and shaky in vague and slowly updated environmental laws. In order to understand China's rules for environmental protection, laws are not the most reliable sources.

Nevertheless, ironically the weak status of rule of law in China further strengthened the decentralization of environmental policy making. Although the National People's Congress is distinctly different from that in a democracy, laws are nevertheless more stable and more authoritative than policies by the administration. Laws are based on wider participation, and the legislative process is more transparent. If strong enough incentives are present, the variety of policy-making entities at different levels will be able to actively innovate new policies, learn the lessons and experiences from other policy making entities, adapt top-down policies and adopt policies from other regional contexts. Not all policy making is necessarily backed by sound research or intellectual support. Nevertheless, the decentralized policy making makes active bottom-up policy innovation and diffusion possible.

5   Decentralized policy implementation

From the perspectives of human resources and fiscal expenditures, China's capacity for environmental policy implementation is heavily tilted toward local governments, rather than the central government.

5.1   *Decentralized human resources*

Policy implementation demands substantially more resources and personnel than policy making. Corresponding to the designed focuses between policy making and implementation, most of China's environmental protection officials are at the municipality and county levels. China had 232,388 government employees on environmental protection in 2015, a 62.8% increase from 142,766 in 2001 to reflect the elevated priority of environmental protection in all government affairs. The distributions across the four levels of governments have been quite consistent over the years, with 1.3%, 6.8%, 21.5% and 63.1% of the total environmental protection personnel in 2015 in central, provincial, municipality and county governments, respectively. Corresponding to the four categories, the municipality and county levels accounted for 92.5% personnel for administration, 97.0% for inspection, 94.6% for monitoring and 69.5% for others (Figure 3.1). As a result, the environmental authorities at the central and even the provincial levels do not have an adequate human resource capacity to implement environmental policies in millions of polluting sources that are scattered in China's wide geographic territories (Ministry of Environmental Protection et al., 2010).

5.2 Decentralized fiscal expenditure and centralized fiscal revenue

Fiscal revenue and expenditure are other key perspectives for understanding the central–local relationship in China. The governmental expenditure-to-GDP ratio in China is not high in comparison to that in developed countries. In 2018, the ratio was 24.5%, in which the central government accounted for 3.6% and local governments 20.9% (Figure 3.2). The ratio dropped significantly from 26.8% to 11.1% from 1980

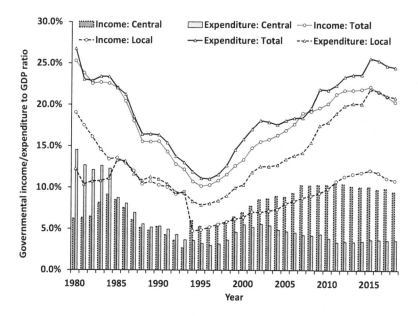

Figure 3.2 Governmental revenue and expenditure to GDP ratios by central and local governments in China

Source: National Bureau of Statistics (2019).

to the mid-1990s but has since gradually recovered (Figure 3.2). The ratio between governmental revenue and GDP had a similar trend, initially falling from 25.3% in 1990 to 10.2% in 1995 and then rising back to 20.4% in 2018 (Figure 3.2). The gaps between revenue and expenditure indicate fiscal surplus or deficit.

In the current fiscal arrangement, the central government has far more revenue than it spends while the local governments in general have to rely on fiscal transfers from the central government for meeting their expenditures. In 2018, the central government received 46.6% of total general fiscal revenue but accounted for only 14.8% of total fiscal expenditures. Local governments, in contrast, received nearly half of the revenue but had to shoulder 85.2% of the expenditures.

The fiscal relationship between the central and local governments have experienced dramatic changes in the past four decades. In 1980, local governments directly received an overall revenue of 87.5 billion RMB (current price), but their spending was 56.2 billion RMB (National Bureau of Statistics, 2019). In contrast, the central government had a revenue of 28.4 billion RMB but spent 66.7 billion RMB. It was the central government, not local governments, that spent most of the government budget, ranging from 52.5% to 55.0% between 1980 and 1984 (Figure 3.2). Accordingly, the central government ran a huge deficit, and local governments, a huge surplus. The fiscal transfer was then from local governments to the central government. It reflected that China's governance remained very much centralized in the immediate years after the Cultural Revolution. The central government was directly engaged in providing a significant proportion of government services and subsidies. Correspondingly, fiscal expenditures were required to support such a provision.

The situation was dramatically changed in 1985. When the governmental expenditure-to-GDP ratio started to drop significantly together with market-oriented economic reforms, the central government saw a much steeper decline (Figure 3.2). The budgets for both the central and local governments became individually more balanced (Figure 3.3). The expenditures of the central and local governments were only 3.3% above and 2.1% lower than their revenues in 1985. Local governments since then have consistently accounted for more than 60% of total governmental expenditures, dwarfing the share of the central government. Although local governments' fiscal conditions remain generally balanced in the following years, the central government again started to see a widening gap. In 1993, its expenditures exceeded revenue by 37.0% while its shares in total government revenue and expenditures had dropped to 22.0% and 28.3%, respectively. The budget deficit of the central government fiscally constrained it from exerting authority on rich provinces and tackling widening regional disparities across the country.

In China's central–local fiscal relationship, 1994 was a crucial watershed when a fundamental tax reform entered into effect in January (State Council, 1993). The central government's share of total governmental revenue skyrocketed to 55.7% in 1994 while its share of expenditures remained at 30.3%. For the first time, the central government ran a budget surplus, with revenue exceeding expenditures by 65.7%. In contrast, local governments' fiscal revenue could cover only 57.2% of

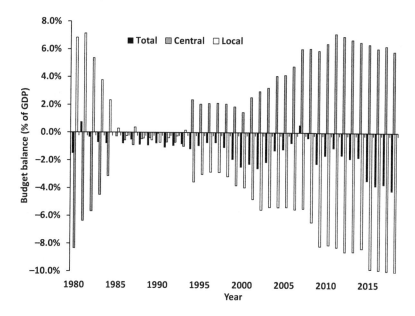

Figure 3.3 Budget balance of central and local governments in China as a proportion of
GDP

Source: National Bureau of Statistics (2019).

their expenditures. Then a large fiscal transfer became necessary from the central
government to local governments. With further decentralization of governmental
affairs and service provision, this newly formed central–local fiscal relationship
has been kept increasingly entrenched in the past two decades. In 2018, local governments accounted for 85.2% of expenditures but only 53.4% of revenue. The
gap has significantly widened.

The current central–local relationship that features significant fiscal transfer
from the central government to local governments reflects their differentiated
roles in policy making and implementation as discussed earlier. The central
government is primarily in charge of policy making while the implementation is largely in the hands of local governments. The former requires much
less expenditure than the latter. All provinces have their expenditures exceeding revenues, but poor provinces tend to rely on the central government's
fiscal transfer much more than rich ones (Figure 3.4). For example, Tibet's
governmental revenue covered only 11.7% of its expenditures in 2018, while
the revenue–expenditure gap for Shanghai was only 14.9%. Accordingly, the
central government could use fiscal transfer as an incentive for local governments to implement policies or achieve goals that are enacted from the top.
It is one of the key incentives that the central government can rely on for the
cooperation of local governments.

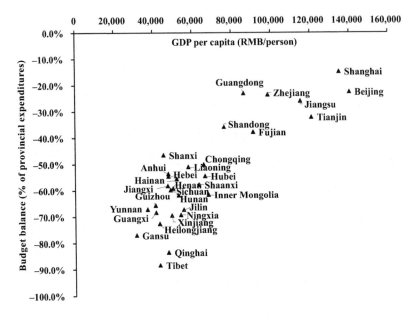

Figure 3.4 Governmental budget balance by provinces as a proportion of governmental expenditures in 2018

Source: National Bureau of Statistics (2019).

With increasing decentralization in the economic reform, more and more budgetary items were shifted with local governments as primary entities of governmental expenditures. Reflecting the division of governmental affairs, the central and local governments now have distinct responsibilities on a variety of expenditure items. Foreign affairs and national defense are two budgetary items that the central government takes almost exclusive responsibility to account for 99.5% and 98.1%, respectively, of total governmental expenditures. Of the central government's expenditure in 2018, 33.8% was devoted to national defense. Grain storage is for the country's food security, and thus, the central government remained more important, being responsible for 66.8% of all governmental expenditures in 2018 (Figure 3.5). Science and technology is another classical category of public good that the market underinvests in to require public expenditures, in which the central government took a share of 37.5% in 2018 (Figure 3.5). Health care and urban and rural communities are almost exclusively the responsibility of local governments. Environmental protection was responsible for 2.9% of total governmental expenditures in 2018 (Figure 3.6), while local governments accounted for 93.2% (Figure 3.5). It occupied 3.1% of local governments' expenditures and 1.3% of the central government's (Figure 3.6).

The expenditure structures between the central and local governments have remained generally unchanged for environmental protection in the past decade. However, this largely decentralized budgetary item has also witnessed signs of

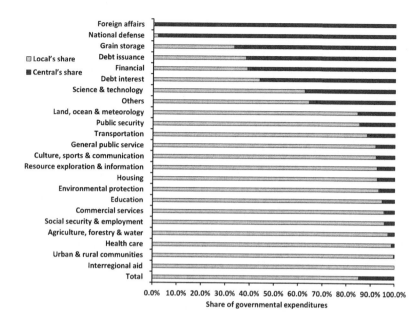

Figure 3.5 The central and local governments' shares of expenditures by budgetary items in 2018

Source: National Bureau of Statistics (2019).

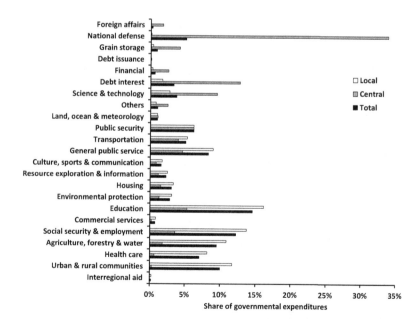

Figure 3.6 Central, local and overall governmental expenditures by budgetary items in 2018

Source: National Bureau of Statistics (2019).

slight recentralization. Recent reforms as described earlier reflected and enabled the central government to be keener in improving environmental quality and more directly involved in supervising local governments. Environmental protection has been listed as a separate budgetary item in the data from the *China Statistical Yearbook* since the 2008 edition (for 2007 data). Its share in total governmental expenditures has inched up from 2.0% in 2007 to 2.7% in 2010 and then fluctuated narrowly to reach 2.9% in 2018. The share in local governments' budgets has also been quite stable, within a narrow range between 2.5% and 3.2% over the period. However, the central government had a significant shift, allocating a much greater share of its budget for environmental protection. It ranged between 0.2% and 0.5% from 2007 to 2013 but then jumped to 1.5% in 2014 and has remained at the level since then (Figure 3.7). Correspondingly, the central government's share in total environmental protection expenditures was lifted from 2.9% in 2013 to 9.0% in 2014, while the local governments' share dropped although their environmental protection expenditures were increased every year in absolute terms.

The significant uplifting in 2014 indicates that environmental protection has been increasingly prioritized in China's public affairs (Figure 3.7). The additional budget mainly corresponded to the strengthened functions of top-down supervision, monitoring and inspection of local governments' performance. Because the shares in governmental expenditure for China as a whole and for local governments

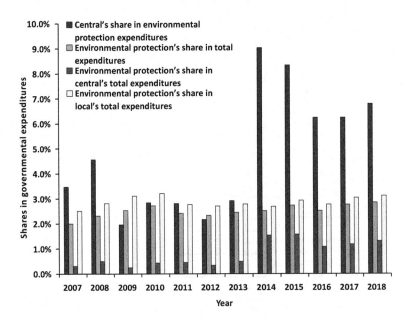

Figure 3.7 Shares in governmental expenditures

Source: National Bureau of Statistics (2019).

Note: The *National Statistical Yearbook* listed environmental protection as a separate budgetary item for the first time in 2007.

did not change significantly over the period and especially in 2014, environmental administrative capacities were not expected to be upgraded disproportionally against other governmental affairs. The emphasis on environmental protection thus targeted the relationship between the central and local governments to more effectively mobilize implementation capacities and to assign a heavier weighting to environmental protection relative to local economic development.

6 Centralized and decentralized personnel management

According to the Chinese Constitution, local leaders are elected by corresponding local People's Congress. Then they are supposed to mainly please their local electorate. Because China's weak rule of law does not ensure the local implementation of environmental laws and policies from the National People's Congress and the central government, the central government should only be able to exert very constrained authority over the selection of local government leaders and what governmental affairs they decide to pursue in their local contexts. Even if local leaders refused to implement policies from the top, only local People's Congress can remove them. However, this very decentralized arrangement presents a sharp contrast with reality. Far-reaching reforms in the past four decades have featured economic reforms on the relationship between the state and the market and administrative reforms on the relationship between the central and local governments, but the relationship between the Chinese Communist Party and the Chinese government has witnessed fewer changes. In the 1980s, their separation was debated and explored in tentative reforms, but the progress has been much slower.

The party plays a crucial role in shaping the central–local leadership relationship in reality. The party and the Chinese government have overlapped organizations in the governmental bureaucracy, while the party is even more prevalent to be present in enterprises and other nongovernmental organizations. Although local leaders should be elected by local People's Congress, the party, and especially its Department of Organization, controls the nominations. For the four governmental levels, each level has the authority to appoint leaders at one lower level. For example, the party's Department of Organization at the central level controls the nomination of provincial-level leaders (including those in central ministries). Each provincial Department of Organization nominates municipality-level leaders within the province. The appointment decisions are in the hands of their corresponding party committees. As a result, personnel decisions are one crucial channel for the central government to influence local governments. Numerous studies have confirmed that China does put governance performance in the decisions to promote or remove officials, especially local government leaders (Li and Zhou, 2005; Zhou, 2007). Without the party's role, China's governance would be substantially different from the current institutional arrangement.

Furthermore, a reform did bring a major change in this personnel relationship with significant decentralization. In 1984, the Central Committee of the party reformed its personnel management system (Gao and Zou, 2007). Before then, the Department of Organization at each level managed two levels down. For example, the Central Department of Organization was in charge of nominating and managing leaders at the provincial and municipality levels. After the reform, the leaders at the municipality level are left to the sole responsibility of the Provincial Department of Organization, while the Central Department of Organization only takes care of the provincial-level leaders. As a result, the provincial leaders will have much stronger control of their staff and other local government leaders below them. Such reform substantially reinforces local leaders' authorities within their jurisdictions. The arrangement coincides with the decentralization of governmental affairs and expenditures but still maintains a powerful channel through the party for the central government to control local leaders.

References

Department of Organization of the Central Committee of the Communist Party of China. 1999. *On reforming the institutions of managing environmental officials*. Beijing, China: Central Committee of the Communist Party of China.

Gao, X. & Zou, Q. 2007. *Research on intra-party democracy – evaluation from history and reality*. Shandong, China: Qingdao Press.

The General Office of the CPC Central Committee & The General Office of the State Council. 2016. *Guiding advice on the pilot vertical reform in sub-provincial monitoring, inspection and enforcement agencies*. Beijing, China: CPC Central Committee, State Council.

Kaufmann, D. & Kraay, A. 2019. *The worldwide governance indicators 2019 update: Aggregate governance indicators 1996–2018* [Online]. Available: https://info.worldbank.org/governance/wgi/.

Li, H. B. & Zhou, L. A. 2005. Political turnover and economic performance: The incentive role of personnel control in China. *Journal of Public Economics*, 89, 1743–1762.

Ministry of Environmental Protection. 2002–2016. *Annual statistical report on the environment in China*. Beijing, China: Ministry of Environmental Protection.

Ministry of Environmental Protection, National Statistics Bureau & Ministry of Agriculture. 2010. *Public report on the first national census of polluting sources*. Beijing, China: Ministry of Environmental Protection, National Statistics Bureau.

National Bureau of Statistics. 2019. *China statistical yearbook*. Beijing, China: China Statistics Press.

National People's Congress. 2000. *Law of atmospheric pollution prevention and control of people's republic of China*. Beijing, China: The 4th Conference of the 10th National People's Congress.

National People's Congress. 2017. *Law of water pollution prevention and control*. Beijing, China: The 4th Conference of the 10th National People's Congress.

National People's Congress. 2018. *Law of atmospheric pollution prevention and control of people's republic of China*. Beijing, China: The 4th Conference of the 10th National People's Congress.

SCOPSR. 2018. *The function, internal organization and personnel of the ministry of ecology and environment*. Beijing, China: SCOPSR.

State Council. 1993. *Decision on implementing the tax sharing mechanism in fiscal management*. Beijing, China: State Council.

State Council. 2018. *Reform plan on the state council*. Beijing, China: State Council.

The U.S. Congress. 1990. *Clean air act amendments 1990*. Washington, DC: The U.S. Congress.

Wang, H., Dong, Z., Xu, Y. & Ge, C. 2016. Eco-compensation for watershed services in China. *Water International*, 41, 271–289.

Zhou, L. 2007. Governing China's local officials: An analysis of promotion tournament model. *Economic Research Journal*, 7, 36–50.

4 Mobilizing the government[1]

1 Goals in China's Five-Year Plans

China's top leadership has gradually gained strong enough political will for environmental protection over the past decades (Chapter 2). However, the decentralization of policy making and, to a greater extent, policy implementation requires the cooperation between the central and local governments to realize the environmental political will with concrete improvement of environmental quality and pollution mitigation (Chapter 3). This chapter is devoted to understanding how the entire Chinese government, from central to local governments, is mobilized through environmental goals, especially in Five-Year Plans.

Goals have been widely used in governance. For example, UNFCCC (United Nations Framework Convention on Climate Change) defines its goal as "stabilization of greenhouse gas concentrations in the atmosphere at a level that would prevent dangerous anthropogenic interference with the climate system" (United Nations, 1992). President Barack Obama set up a goal to withdraw all U.S. troops from Iraq by the end of 2011 (DeYoung, February 28, 2009). Many studies are about environmental goals, including those on negotiating goals, distributing goals (Chakravarty et al., 2009), policies to achieve goals (such as on emission tax and cap-and-trade) and technological achievability of goals (Pacala and Socolow, 2004).

A theoretical foundation of using goals as a governance tool can be traced to studies in social psychology: through experiments on individuals, the impact of various goals on task performance is examined. Locke et al. (1981) reviewed the literature and concluded that "specific and challenging goals lead to higher performance than easy goals, 'do your best' goals, or no goals." Furthermore,

> goal setting is most likely to improve task performance when . . . the subjects have sufficient ability, . . . feedback is provided to show progress in relation to the goal, rewards such as money are given for goal attainment, the experimenter or manager is supportive, and assigned goals are accepted by the individual.
>
> (Locke et al., 1981)

In the experiments, goals are distributed to individuals and individuals try to accomplish the goals. The situation is not much different from an environmental goal in a big country like China. The Chinese central government plays a similar role as experimenters: it decides a goal and distributes it to local governments. Three components could be distinguished: (1) goal setting, (2) goal distribution and (3) goal attainment. Goal setting refers to what type of goals should be set up and how stringent they are. Because a global or national goal often requires the cooperation of different political or administrative entities, goal distribution is necessary. For example, a global goal of carbon dioxide (CO_2) mitigation should be distributed to individual countries, and a Chinese national goal should be distributed to provinces. Furthermore, these goals need to be accepted before serious efforts are made. The third component of a goal process focuses on evaluating goal attainment. Strong-enough incentives should be put into place to mobilize goal implementers.

The seven-decade history of the People's Republic of China can be divided into two periods: a centrally planned economy in the first three decades and a later era of market-oriented economic reforms. Since the 1950s, originally adopted from the Soviet Union, Five-Year Plans have become pivotal to guide China's economic development. Although China's economy was strictly state-controlled before the economic reforms began in 1978, only the first of the earliest five Five-Year Plans was actually completed (Liu et al., 2006). The other four were not able to be performed due to frequent political movements, with the Cultural Revolution as the most notable one (Liu et al., 2006).

Five-Year Plans gained momentum only in the second period when China tried to establish a market-oriented economy. Starting from the 6th Five-Year Plan (1981–1985), China has gradually formed a set of rules to design these plans (State Council, 2005b). The 11th Five-Year Plan (2006–2010) was the first to change its name from "*jihua*" (more forceful plans) to "*guihua*" (more directional plans). Goals are the most important indicators in the Plans. From the 11th Five-Year Plan, goals are distinguished into foreseeable ones (such as the growth rates of gross domestic product [GDP] and population) and legally binding ones (such as pollutant mitigation; National People's Congress, 2006). In addition, China's Five-Year Plans are not just one document but a system composed of many layers. For example, for the nation as a whole, there was a National 11th Five-Year Plan that included a 10% reduction goal of sulfur dioxide (SO_2) emissions. Another 11th Five-Year Plan on Environmental Protection provided further details. At one more layer lower, the 11th Five-Year Plan on Acid Rain and SO_2 Pollution Control specifically addressed the mitigation of SO_2 emissions. There were also 11th Five-Year Plans at all governmental levels.

Goals are playing more and more prominent roles in China's environmental protection, especially in Five-Year Plans, to mobilize local governments and the Chinese bureaucracy. If expressed in percentage terms, the baseline year is the final year of the previous Five-Year Plan. For example, China's energy intensity goal in the 11th Five-Year Plan (2006–2010) was a 20% reduction (National People's Congress, 2006); it indicates that China planned to reduce energy intensity, or energy consumption per unit of GDP, by 20% in 2010 from the 2005 level.

In regulating SO_2 emissions that mainly come from the burning of coal, China relies on absolute emission goals, which were a 3.8% increase, a 10% reduction, a 10% reduction, an 8% reduction and a 15% reduction, respectively, for the 9th, 10th, 11th, 12th and 13th Five-Year Plans (National People's Congress, 2001, 2006, 2011; NEPA et al., 1996; National People's Congress, 2016). The actual growth rates of SO_2 emissions were a 15.8% reduction, a 27.8% increase, a 14.3% reduction and a 14.9% reduction, respective for the 9th, 10th, 11th and 12th Five-Year Plans, indicating goal attainment in all but the 10th Five-Year Plan (National Statistics Bureau and Ministry of Ecology and Environment, 2019). This chapter specifically analyzes the 10% reduction goal of SO_2 emissions in the 11th Five-Year Plan as it reversed the humiliating failure in the 10th Five-Year Plan. The national quantitative goal was centrally set up to involve the Chinese top leadership and the then State Environmental Protection Administration (SEPA, presently the Ministry of Ecology and Environment). The mitigation tasks were distributed to provincial and other local governments with their individual goals. Mechanisms were put into place to monitor the goal compliance statuses of local governments and take enforcement actions for their cooperation. Goals have also been rapidly evolving to reflect the status and intended emphasis of SO_2 mitigation and air pollution control.

2 Centralized goal setting

2.1 Setting up the national goal

China's goal process involves three overlapping cycles: Five-Year Plans, National Party's Congresses and National People's Congresses. The 11th Five-Year Plan formally started in 2006 and concluded in 2010. The 16th National Party's Congress lasted from October 2002 to October 2007. The 10th National People's Congress lagged half a year behind, from March 2003 to March 2008. The 11th Five-Year Plan did not begin until the middle of the two Congresses. Under China's present political reality, the two Congresses have a reasonable sequence. The National Party's Congress selects party leaders. After a further distribution of power, these leaders assume various governmental jobs in the following National People's Congress. The first gatherings of these two Congresses are mainly about determining the leadership of the party and the country. Then China's leaders reshuffle every five years. As a result, the three cycles are actually two: the Five-Year Plans and the change of leadership.

The cycles have existed in the present form for about four decades, especially since 1992. The most stable cycle is the Five-Year Plan. All Five-Year Plans are targeted for five years, even in the most irrational period of the Cultural Revolution. Since the 3rd Five-Year Plan (1966–1970), the period has been consecutive. The National Party's Congress formed its own five-year cycle in 1977, and the National People's Congress, in 1978. But the leadership change did not match the Congresses' cycles until 14 years later. Jiang Zemin was formally elected as the secretary general of the party in 1992 and the president of China in 1993. Since then, China's top leaders also have established their five-year cycles, formally synchronized with the Congresses.

The cycles of Five-Year Plans do not match China's change of leadership. The anchor year of a Five-Year Plan is the previous year before the plan starts. But because the plan has to be formed before all information in the anchor year is known and China's SO_2 emissions are very volatile, relative goals are much better than absolute goals to address the huge uncertainty. China's failure to attain the 10% reduction goal of SO_2 emissions in the 10th Five-Year Plan (2001–2005) may partly reflect the mismatch among cycles. A new administration took full charge in March 2003 when the 10th Five-Year Plan had been going on for over two years. Almost immediately afterward, China's SO_2 emissions went out of control. During 2001–2002, SO_2 emissions went down by 3.4%, but in the remaining three years (2003–2005), the emissions surged by 32.3% (SEPA, 2001–2009). On the other hand, the sharp contrast was not obvious from the perspective of economic growth. In annual terms, China's economy expanded at an annual rate of 8.7% in the first two years and 10.2% later (National Bureau of Statistics of China, 1999). Although the surge could be simply a coincidence with the change of leadership, if 2003 through 2005 had been under the same administration as in 2001–2002, the result might be different due to a better unification of planning and implementation.

The Outline of the National 11th Five-Year Plan on Economic and Social Development (hereafter referred to as the *Outline*) was the title of an official document ratified by the National People's Congress, the nominally highest authority in China, in March 2006 (National People's Congress, 2006). The 10% reduction goal of SO_2 emissions was clearly included to be legally binding. The process to reach the *Outline* can be divided into three periods: (1) mid-2003 to December 2004, concluded with the formation of *The Basic Thoughts of the National 11th Five-Year Plan* (hereafter referred to as the *Basic Thoughts*; National Development and Reform Commissions, or NDRC, was responsible); (2) February 2005 to October 2005, ended with the ratification of *The Suggestions on Designing the National 11th Five-Year Plan* (hereafter referred to as the *Suggestions*; the Central Committee of the Chinese Communist Party was in charge); (3) October 2005 to March 2006, indicated by the enactment of the *Outline* (State Council took the hold).

The *Basic Thoughts* contemplated the strategic direction of the *Outline*. This idea-framing period was initiated in mid-2003 and completed by the end of 2004 (Xinhua News Agency, 2006; NDRC, 2003). For environmental protection, the job of the 11th Five-Year Plan was to "decelerate the trend of ecological and environmental deterioration and strengthen the ability of sustainable development" (NDRC, 2005). The wording clearly differs from, for example, "improving environmental quality." It may be reflected later in the *Basic Thoughts on Environmental Protection* with a flat SO_2 emission goal proposed (SEPA, 2006d; Chinese Academy for Environmental Planning [CAEP], 2004).

The then named SEPA was responsible for writing the 11th Five-Year Plan for Environmental Protection. The SEPA understood the specific difficulty of controlling SO_2 emissions. For example, in 2002, Wang Xinfang, a deputy administrator of the SEPA, admitted that it was hard to achieve the 10% reduction goal of SO_2

emissions in the 10th Five-Year Plan (2001–2005; Wang, 2002). The final result in 2005 confirmed his concern: goals on other pollutants were either met or slightly missed, but SO_2 emissions were 27.8% higher than the level in 2000 and 42% higher than the original goal (Zou et al., 2006). The SEPA distributed *The Basic Thoughts on Environmental Protection* on December 23, 2004, and proposed a flat goal for the 11th Five-Year Plan (SEPA, 2006d; CAEP, 2004). The midterm assessment on the 10th Five-Year Plan that was completed in 2004 could have played a guiding role in the proposal: the available data showed an 8.2% increase of SO_2 emissions in 2003 compared with those in 2000 (SEPA, 2001–2009). The midterm assessment believed that the 10% reduction goal had fallen out of reach but still expected that SO_2 emissions in 2005 could remain the same as the level in 2000 (Zou et al., 2004).

With the tentative *Basic Thoughts*, the top leadership in the Central Committee of the Chinese Communist Party got directly involved. The period was formally initiated with the establishment of a high-profile drafting team on February 16, 2005, headed directly by Premier Wen Jiabao (Xinhua News Agency, 2005). A prominent feature is the multiple meetings presided by President Hu Jintao in the Political Bureau or its Standing Committee and by Premier Wen Jiabao in the drafting team (Xinhua News Agency, 2005). The *Suggestions* was finally passed and endorsed on October 11, 2005, by the Central Committee of the Chinese Communist Party (Xinhua News Agency, 2005). Sharply different from the *Basic Thoughts*, the *Suggestions* clearly declared to "reduce total emissions of pollutants," which essentially indicated a goal of improving environmental quality (Xinhua News Agency, 2005).

After the *Suggestions* tightened the goal for environmental protection in October 2005, the third period started with the establishment of a drafting team that comprised various ministries in the central government (Xinhua News Agency, 2006). An expert committee was summoned to comment on the drafts of the *Outline* (Ma, 2005). The public was also consulted for advice (Ma, 2005). President Hu Jintao and Premier Wen Jiabao organized several meetings to discuss the drafts (Xinhua News Agency, 2006). In November 2005, the SEPA drafted a plan on acid rain and SO_2 emission control (SEPA, 2005). Although SO_2 emissions in 2004 had been 13% higher than the 2000 level, the 10% reduction goal for the 11th Five-Year Plan first appeared (SEPA, 2005, 2001–2009). On December 3, 2005, State Council enacted *Decisions on Realizing Scientific View of Development and Strengthening Environmental Protection* (State Council, 2005a), which linked the new ideology of Scientific View of Development with environmental protection. It confirmed the importance of environmental protection in the establishment of the new ideology. When the 4th Conference of the 10th National People's Congress was in session, the *Outline* was submitted on March 5, 2006, and approved on March 14, 2006 (Xinhua News Agency, 2006).

2.2 Methods of goal setting

A Five-Year Plan anchors at the previous year of its planning period. For example, a goal in the 11th Five-Year Plan (2006–2010) is to compare 2010 with 2005. In practice, the anchor year's data cannot be fully utilized in setting up the goals. China generally published environmental data for the previous year in around June (SEPA, 2001–2009). Although the public may get the information later than the Chinese government, several months could elapse for the collection and compilation of data. Accordingly, the anchor year's information cannot be fully employed in planning but has to be the foundation for the next Five-Year Plan. China's annual change of SO_2 emissions varied greatly: the 2004 emissions were 4.5% up from the 2003 level, but the figure surprisingly jumped 13.1% in 2005 (SEPA, 2001–2009). At the same time, however, the economic growth rates were quite stable with 10.1% in 2004 and 11.4% in 2005 (National Bureau of Statistics, 2019). Because of the substantial volatility, the absence of data in the most relevant and important anchor year could cause significant trouble in calibrating goals.

Two components were important in setting up China's SO_2 emission goals in Five-Year Plans: long-term goals and appropriate mitigation paces. China relied on a concept called "environmental capacity" to decide long-term SO_2 emission goals (Yang et al., 1998, 1999). "Environmental capacity" refers to the upper-limit emissions of a pollutant without degrading a kind of environmental quality below a minimum level. The environmental capacity for SO_2 emissions is a function of three variables: (1) the amount and distribution of SO_2 emissions, or emission inventories; (2) the transport and sinks of SO_2; and (3) an acceptable level of some environmental quality. The second variable is largely determined by atmospheric circulation and chemistry. The third variable was used as an external choice. If society would like to live in a better environment, the limit of ambient SO_2 concentration could be lowered and SO_2 emissions have to be further reduced.

To set up an SO_2 goal in a Five-Year Plan, China first decided on a long-term goal and then found an appropriate mitigation pace to attain the goal. The long-term goals were determined with models of atmospheric transport and chemistry. The implicit long-term goal for the 10th Five-Year Plan (2001–2005) was 12 million tons and was scheduled to get attained in 2020 (Wang, 2002). For the 11th Five-Year Plan (2006–2010), the long-term goal became 18 million tons and the goal attainment year would also be 2020 (SEPA, 2005). Although both goals were supported by scientific research with different constraint conditions, the significant upward revision of the long-term goal probably arose as a result of the sharp increase in coal use that led to an unanticipated rise of SO_2 emissions in the 10th Five-Year Plan.

The long-term goals have certain scientific foundations. China's Law of Environmental Protection clearly holds local governments responsible for local environmental quality (National People's Congress, 1989). Because ambient air quality standards are also "mandatory standards" in the Law of Standardization (State Council, 1990), local government leaders should be mobilized to enforce SO_2 mitigation policies if the law were well respected. In 1996, the then State Environmental Protection Agency enacted ambient air quality standards (NEPA and SBTS, 1996). Most of China's land area with economic and human activities should have ambient SO_2 concentration in annual mean below 0.060 mg/m^3. One key study showed that only to achieve this average concentration within grid boxes of

$0.2° \times 0.2°$, China has to control its SO_2 emissions at 12 million tons (Yang et al., 1999). Another study for the 11th Five-Year Plan selected critical acid deposition within grid boxes of $1° \times 1°$ (Zou et al., 2006). Although the number was based on several heavy assumptions (most important, the geographical distribution of SO_2 emission sources), it signaled the stringency of the ambient SO_2 concentration standard. For example, China's goal in the 11th Five-Year Plan was to reduce SO_2 emissions from 25.5 million tons in 2005 by 10% in 2010, still far above the 12-million-ton level (National People's Congress, 2006).

The distribution of SO_2 emissions matters greatly for any national SO_2 mitigation goal that is based on SO_2 concentration. For example, with SO_2 concentration of 0.060 mg/m^3 as the constraint condition, Shanghai could emit up to 0.63 million tons of SO_2 (Yang et al., 1999), but its actual emissions in 2007 were 0.50 million tons (Ministry of Environmental Protection, 2008). Then if a polluting source was located in Shanghai, it would have no necessity to mitigate. But if the same source were moved to Jiangsu, a neighboring province with its emission limit below actual emissions (Ministry of Environmental Protection, 2008; Yang et al., 1999), it would be subject to serious abatement. The 1998 study revealed a goal based on SO_2 ambient concentration: if not counting the excess environmental capacity in Tibet compared with its emissions (0.50 million tons vs. 1.5 thousand tons), China's national goal was to reduce SO_2 emissions to about 12 million tons (Yang et al., 1999). China planned to attain the goal in 2020 (Wang, 2002). The goal for the 10th Five-Year Plan was then established as a 10% reduction, or 18 million tons (SEPA, 2001).

After the big failure in the 10th Five-Year Plan on SO_2 mitigation, China still held 2020 as the attainment year of a long-term goal. However, the original goal would be too difficult. In 2005, China emitted 25.5 million tons of SO_2 (SEPA, 2001–2009). To achieve the goal of 12 million tons in 2020, a 53% reduction in 15 years would be required. Even if from the 2004 level when a new goal for the 11th Five-Year Plan was formed, the reduction rate should still be 47% (SEPA, 2001–2009). By replacing the constraints of SO_2 concentration with critical acid deposition, a new environmental capacity was worked out to be 17.3 million tons (Zou et al., 2006). Then 18 million tons were chosen to be the new long-term goal (SEPA, 2005). These two long-term goals assumed a similar pace of about 2 to 2.5 million tons reduction per five years. Because of the relatively stable pace and a common attainment year of the long-term goals, China's long-term goals seemed to be reversely decided from current emission levels. Interestingly, both long-term goals were supported by scientific research. The history could indicate that the results of the scientific research were selected beforehand by nonscientific factors.

In deciding goals for Five-Year Plans, the emission trends in previous years were also considered (Wang et al., 2004). Because the 9th Five-Year Plan achieved a 15.8% reduction (NEPA et al., 1996; SEPA, 2001–2009), even a similar trend was thought to be too stringent (Wang et al., 2004). Probably the 10% reduction goal was established because it stood between the 15.8% reduction and the original goal of a 3.8% increase in the 9th Five-Year Plan. A middle ground, closer to the stringent end, was taken. On the other hand, the same historical trend would be too relaxed for the 11th Five-Year Plan. SO_2 emissions went up by 27.8% in the 10th Five-Year Plan (SEPA, 2001–2009). Certainly this was not an acceptable trend, but it might be an important factor that drove the initial flat goal for the 11th Five-Year Plan (CAEP, 2004). The same principle could have been followed: 0% change was closer to the stringent end between a 27.8% increase and a 10% reduction. As a result, the goal attainment in the previous Five-Year Plan should have played an important role in framing a goal for the next.

The United States' goal of SO_2 emissions in Clean Air Act Amendments (CAAA; 1990) was also expressed in relative terms. Relative to the emission level in the anchor year of 1980, SO_2 emissions were planned for reduction by 10 million tons (The U.S. Congress, 1990). Although an intensive 10-year study was performed in the 1980s (National Acid Precipitation Assessment Program), it failed to answer relevant questions for policy making and was not closely connected to the goal-setting process (Roberts, 1991; Pouyat and McGlinch, 1998). For a fixed long-term goal, different anchor years only correspond to different relative reductions or different expressions of the figures. Furthermore, 1980 was not a baseline year for allowance allocation. Rather, 1985 was a much more important year with real implications in grandfathering emission permits. However, if the 10-million-ton reduction was fixed, the choice of 1980 did have important implications. In 1980, the U.S. emitted 23.5 million tons of SO_2 and the figures in 1985 and 1990 were, respectively, 21.1 and 20.9 million tons (U.S. Environmental Protection Agency, 2007). Essentially, the choice of 1985 and 1990 would have no difference. But anchoring in 1980 could effectively relax the long-term goal by about 2.4 to 2.6 million tons. The goal was planned for attainment in 2010. The anchor year 1980 was ten years ahead of the legislation and 15 years before the program formally started in 1995. Although whether a goal was expressed in relative or absolute terms matters greatly in China, it was generally not quite relevant for the United States' goal setting. The United States had much less volatility in annual SO_2 emissions. The burden to achieve the goal – the difference between business-as-usual emissions and the goal – was accordingly much less uncertain than China's. The major benefit of relative terms was to reduce the uncertainty of surprising emission growth or reduction. However, less uncertainty in the United States and the longer goal cycle did not distinguish this benefit. Furthermore, the Acid Rain Program's goal cycle was much longer than China's Five-Year Plans. Because of the well-established rule of law, the law ensured that the SO_2 mitigation efforts would continue regardless of who was the president or which political party he or she belonged to.

3 Top-down goal distribution

Goal implementation refers to a process for goal implementers to receive, accept and work for goal attainment. It is quite different from policy implementation. Goal implementation deals with the relationship among different governments or their agencies, while policy implementation focuses on the relationship between the government and polluters, including industrial plants and individuals. Goal setters and goal implementers are usually separate in the Chinese government. Since goal setters are not directly in charge of achieving the goal, they have to find a way to get goal implementers to accept the goal and to work hard for it. In order for effective goal implementation, subgoals should be created from the national goal to demand an appropriate distribution scheme. The UNFCCC defines a principle of sharing the duty of reducing greenhouse gas emissions among countries according to "common but differentiated responsibilities and respective capabilities" (United Nations, 1992). Which applicable principles should be followed has attracted negotiation debates and academic studies (Chakravarty et al., 2009; Li, 2010).

A national goal and its distribution to local governments often fall into separate decision-making processes in the Chinese setting. Taking the SO_2 goal in the 11th Five-Year Plan (2006–2010) as an example, the national 10% reduction goal was largely decided by the top leadership of the party, but provincial goals came from a bargaining process between the central government – mainly the then SEPA – and provincial governments.

3.1 Goal distribution from the central to provincial governments

Chinese local governments are divided into several levels, mainly provinces, municipalities and counties. To implement SO_2 emission goals, the central government distributed subgoals to provincial and local governments and issued incentives to mobilize their leaders. A good national goal is hard to implement without a fair distribution of the burden. After a national goal is framed, provinces will negotiate with the central government for their shares of the burden. The details of the negotiation and their applied principles are not publicly available but could be reversely examined from the outcome.

China qualitatively disclosed principles to distribute the national goal to 31 provinces. Key influential factors included environmental quality, environmental capacity, current emission level, economic development status, SO_2 mitigation capability, requirements of various pollution control plans and regional category (west, middle, east; State Council, 2006). An explicit formula was less likely to exist that connected these factors with a province's goal. However, published provincial information could at least lead to an evaluation of potentially quantitative relationships. Econometric analysis was applied here with a linear assumption.

The dependent variable was provincial SO_2 emission goals in percentage terms: SO_2 emission target in 2010 / SO_2 emissions in 2005 – 100%. The distributed national goal, 11.9% reduction, was actually a little more stringent than a 10% reduction (State Council, 2006). All provinces combined could only emit 22.47 million tons, not 22.94 million tons for the nation. The difference (0.47 million tons) was reserved for experimenting with SO_2 emission cap-and-trade (State Council, 2006).

Independent variables included all those factors indicated by the Chinese government (State Council, 2006). Because 2005 was the anchor year of the 11th Five-Year

Plan, independent variables all referred to this year unless otherwise specified. Environmental quality was represented by both the annually average SO_2 concentration in provincial capitals and nonpower sectors' emission density (expressed in tons/km^2). The capitals' SO_2 concentration data were published in *China Statistical Yearbooks* (National Bureau of Statistics, 2006). In addition, China divided SO_2 emissions into two big categories: power and nonpower. Associated with shorter chimneys, non-power-sector emissions were believed to be more closely associated with local air quality (SEPA, 2006a). Their emission densities in provinces were employed to represent another perspective of environmental quality (National Bureau of Statistics of China, 1999; Zou et al., 2006). *Environmental capacity* is a term indicating allowed maximum emissions to maintain a certain environmental quality. The data used in this section came from a study that calculated long-term SO_2 goals for the 11th Five-Year Plan (Zou et al., 2006). Critical acid deposition was the targeted environmental quality. The corresponding upper-limit national emissions were 17.3 million tons, and each province had its own figure (Zou et al., 2006). Current emission levels were represented by provincial SO_2 emissions in 2005. Provincial goals were formally distributed in August 2006 (State Council, 2006). Because data for 2005 had been published in June 2006 (SEPA, 2001–2009), they should be available for negotiating the goal distribution. Provincial GDP per capita stood for economic development status (National Bureau of Statistics, 2006).

No definition had been clearly displayed by the Chinese authorities on SO_2 mitigation capability. Two variables were used. First, higher provincial SO_2 removal rates in 2005 could indicate fewer opportunities for the future. From another aspect, they also represented previous efforts in SO_2 mitigation. Second, SO_2 scrubbers (or flue-gas desulfurization facilities, FGD) had been designated as a key measure to reduce SO_2 emissions in the 11th Five-Year Plan (State Council, 2007a). The power sector's shares of total emissions would then serve as another indicator of mitigation capability (National Bureau of Statistics of China, 1999; Zou et al., 2006). Higher shares may lead to a more effective reduction of total SO_2 emissions through SO_2 scrubbers.

China's policies and emission control plans targeting individual emission sources could decide provincial goals in a bottom-up way. Nevertheless, it may not coincide with the top-down results. For example, effluent emission standards and SO_2 scrubber planning, respectively, were expected to lead to national power sector's emissions of 8.9 and 9.7 million tons in 2010, while the finally assigned goal was 9.5 million tons in the 11th Five-Year Plan (Zou et al., 2006). To evaluate their impact on goal distribution, two independent variables were generated for each province: (1) (Emission standard-designated levels in the power sector in 2010 + Nonpower emission goals in 2010) / Provincial emissions in 2005 – 100% (State Council, 2006; Zou et al., 2006; National Bureau of Statistics of China, 1999) and (2) (Scrubber planning-projected emissions in power sector + Nonpower emission goals in 2010) / Provincial emissions in 2005 – 100% (State Council, 2006; Zou et al., 2006; National Bureau of Statistics of China, 1999).

According to geographical locations and economic advancement, China divides its provinces into three regional groups: west, center and east. To alleviate regional disparity in economic growth and income, China treats the three categories differently. For example, "Great West Development" aimed to develop western provinces, particularly through building infrastructure. Dummy variables were generated to indicate a province's location. In addition, because China's prevalent wind generally transports air pollutants from the west to the east, SO_2 emissions in western provinces could cause more damage than those in eastern provinces. The dummy variables then evaluated the overall impacts of these two opposite concerns.

Besides these variables, several others that were not mentioned in the official distribution plan were also tested, including SO_2 emissions per capita, goal attainment in the 10th Five-Year Plan and electricity export. One argument for China not to accept a legally binding goal on carbon mitigation in the Kyoto Protocol was its low carbon emissions per capita. Whether China applied this principle in domestic practice was examined through provincial SO_2 emissions per capita in 2005 (National Bureau of Statistics of China, 1999).

China failed substantially to achieve its 10% reduction goal of SO_2 emissions in the 10th Five-Year Plan (2001–2005): the actual emissions in 2005 were 42% higher than the original goal (SEPA, 2001–2009, 2001). But some provinces did better than others. Whether better performance in the past was recognized is tested through a ratio: Provincial emissions in 2005 / Provincial emission targets in the 10th Five-Year Plan for 2005 (State Council, 2006; National Bureau of Statistics of China, 1999). In addition, for the 27 provinces used in models (discussed later), this variable was highly correlated with the provincial growth rates of SO_2 emissions in the 10th Five-Year Plan and the correlation coefficient is 0.98. Accordingly, the model results on this goal attainment variable could be almost identically applied to a variable on the growth rates.

Pollutant emissions and product consumption are not necessarily in the same location. Electricity is a clear and important case. SO_2 comes out of coal-fired power plants, but electricity could be lighting bulbs in another province. This effect was examined through provincial electricity trade: Provincial electricity generation / Provincial electricity consumption − 100% (National Bureau of Statistics, 1997–2008).

Although mainland China has 31 provinces, only 27 were used for the statistical models. Four provinces were kept out. Hainan and Tibet had too-insignificant SO_2 emissions in 2005, respectively, 22,000 and 2,000 tons. Qinghai had the least emissions among provinces except the two previously mentioned, and its data on avoided industrial emissions were not available in *China Statistical Yearbooks*. Shanghai had its nonpower SO_2 emission density in 2005 much higher than other provinces (32.7 tons/km²; the next highest was 7.9 tons/km²), a far outlier (National Bureau of Statistics of China, 1999; Zou et al., 2006).

The correlation coefficients between the variables are given in Table 4.1. Provincial goals were highly correlated negatively with nonpower emission density, total SO_2 emissions and GDP per capita – indicating that higher levels of these variables were closely associated with more stringent provincial goals – and

Table 4.1(a) Correlation coefficients of key factors for 27 provinces

	Reduction goal	Capital's SO₂ conc. In 2005	Nonpower emission density in 2005	Long-term goal	Total emissions in 2005	GDP per capita in 2005	SO₂ removal Rate in 2005	Power's emission share in 2005	Scrubber planning decided goals	Emission standard decided goals	Middle	West	SO₂ emission per capita in 2005	Goal attainment in the 10th Five-Year Plan	Electricity export in 2005
Reduction goal	1.00														
Capital's SO₂ concentration	0.00	1.00													
Nonpower emission density	-0.74	0.17	1.00												
Long-term goal	-0.06	0.05	-0.04	1.00											
Total emissions	-0.53	0.16	0.22	0.15	1.00										
GDP/capita	-0.48	-0.04	0.59	-0.30	-0.16	1.00									
Removal rate	0.18	-0.12	-0.01	-0.08	-0.16	-0.13	1.00								
Power's emission share	-0.10	-0.34	0.00	-0.35	0.05	0.25	-0.22	1.00							
Scrubber planning	0.63	0.11	-0.36	0.31	-0.39	-0.33	-0.03	-0.23	1.00						
Emission standard	-0.06	-0.17	0.29	-0.19	-0.41	0.75	0.10	0.10	0.02	1.00					
Middle	0.35	-0.23	-0.32	-0.18	0.01	-0.28	0.05	0.21	0.16	-0.08	1.00				
West	0.24	0.39	-0.26	0.14	-0.17	-0.46	-0.08	-0.29	0.21	-0.54	-0.46	1.00			
Emission/capita	0.01	0.24	0.03	-0.01	0.24	-0.14	-0.40	0.18	-0.03	-0.44	-0.03	0.25	1.00		
Goal attainment	0.44	-0.30	-0.60	-0.05	0.00	-0.22	-0.14	0.30	0.32	-0.07	0.42	-0.17	-0.17	1.00	
Electricity export	0.40	-0.05	-0.47	-0.02	0.21	-0.54	-0.05	0.28	0.10	-0.44	0.52	0.06	0.35	0.31	1.00

Note: China's mainland has 31 provincial regions. Four are not included here: Qinghai, Hainan, Tibet, and Shanghai.

Table 4.1(b) Summary of variables

Variables	Unit	Period	No. of observations	Mean	Std. Dev.	Min	Max
Reduction goal	%	11th Five-Year Plan	27	-10.1	5.7	-20.4	0.0
Capital's SO$_2$ conc.	mg/m^3	2005	27	0.057	0.020	0.02	0.12
Nonpower emission density	ton/km^2	2005	27	2.97	2.09	0.2	7.9
long-term goal	%	11th Five-Year Plan	27	0.81	1.02	-0.3	3.7
Total emissions	10,000 tons	2005	27	91.97	48.39	19.0	200.2
GDP/capita	10,000 RMB/person	2005	27	1.55	0.91	0.5	4.5
Removal rate	%	2005	27	0.28	0.16	0.1	0.6
Power's emission share	%	2005	27	0.52	0.11	0.3	0.7
Scrubber planning	%	11th Five-Year Plan	27	-0.09	0.11	-0.3	0.1
Emission standard	%	11th Five-Year Plan	27	-0.13	0.09	-0.2	0.0
Middle	dummy		27	0.33	0.48	0	1
West	dummy		27	0.30	0.47	0	1
Emission/capita	kg/person	2005	27	22.95	13.26	9.3	61.0
goal attainment	%	10th Five-Year Plan	27	0.51	0.34	0.0	1.3
Electricity export	%	2005	27	0.03	0.24	-0.6	0.6

positively with SO_2 emissions from scrubber planning, goal attainment in the 10th Five-Year Plan and electricity export.

Although nonlinear terms could show consistent significance in models, such as the squared term of nonpower SO_2 emissions density, its actual application in the negotiation for provincial goals was difficult. China very likely did not use a written formula to decide provincial goals. The nonlinear relationship was thus too complicated for arguments, especially those with a turning point. In addition, once included, several far points could greatly change the overall relationship in models. For example, if Shanghai appeared in the models, its big nonpower SO_2 emission density would make the corresponding coefficient much different. These provinces might experience special negotiation. As a result, only linear terms were used in the models to examine China's principles in goal distribution.

Another decision about the regression models was whether a constant variable should be included. If all provinces had to presume a basic reduction goal and adjust it according to specific situations, the constant variable would show significance and the explained variance, R^2, should be higher compared with a no-constant model. Model runs indicated otherwise (Table 4.2). In response, no constant variable appeared in the remaining models.

The model results showed that two variables were the most important in distributing the national goal to provinces. First, richer provinces tended to receive more stringent reduction goals. Provincial GDP per capita in 2005 and provincial goals in the 11th Five-Year Plan had a correlation coefficient of -0.48 (Table 4.1). But statistical models did not consistently show the significance of GDP per capita (Table 4.2). However, if either nonpower emissions density in 2005 or provincial goals from scrubber planning were excluded, GDP per capita would become significant. For every 10,000 RMB/person increase, the province should reduce its SO_2 emissions by further 1.3% (from the 2005 level). In explaining the model results, a problem was that two factors had a high correlation, and both showed significance on some occasions. However, it should not have mattered much in the negotiation. As long as no clear formula decided goals, a province could always argue with one factor to generate a more favorable goal. For example, Shanghai had a much higher nonpower SO_2 emission density in 2005 than other provinces, but its GDP per capita in 2005 was ahead, with a significantly narrower margin (52,000 RMB/person compared with the next highest 45,000; National Bureau of Statistics of China, 1999; Zou et al., 2006). Comparatively, Shanghai could ask for a less stringent goal from GDP-per-capita point of view. Second, provinces with large emissions had tougher goals. China's big provinces experienced greater pressure to reduce their emissions more for achieving the national goal. Coefficients of provincial emissions in 2005 were consistently significant (Table 4.2). Every 100,000 tons more SO_2 emissions corresponded to about 0.47% further reduction. Third, nonpower emissions density displayed consistent significance. For emitting one more ton per square kilometer, a province should further reduce total SO_2 emission by about 1.3%.

Notably, several other variables did not show much influence. First, provinces with worse environmental quality might not have received more stringent goals.

Table 4.2 Regression model results for distributing the national goal to provinces

Independent variables	Model 1	Model 2	Model 3	Model 4	Model 5	Model 6	Model 7	Model 8	Model 9
Capital's SO$_2$ concentration	39.51	40.03	47.28						
Nonpower emission density	-1.14*	-1.14*	-1.22***	-0.80	-1.00*	-0.83*	-1.34***	-1.29***	
Long-term goal	-0.88	-0.86	-0.99	-0.43					
Total emissions	-0.041*	-0.041*	-0.038*	-0.064***	-0.042**	-0.066***	-0.047***	-0.033**	-0.058***
GDP per capita	-1.98	-1.95	-2.25	-1.87*	-1.47	-1.73*	-1.27	-0.97	-2.79***
Removal rate	4.36	4.48	2.40						
Power's emission share	-2.63	-2.42	1.43						
Scrubber planning	14.08	14.10*	15.35**		12.98*			15.39**	10.09
Emission standard	15.47	15.38	9.83		12.55			3.33	12.71
Middle	-0.18	-0.13	0.36						
West	-0.95	-0.90	-0.99						
Emission/capita	0.07	0.07		0.01	0.05	1.98			
Goal attainment	0.57	0.58		2.16	0.73	4.48			3.62**
Electricity export	3.74	3.69		3.96	4.43				6.33**
Constant	0.31								
Adjusted R^2	0.76	0.95	0.94	0.93	0.94	0.93	0.93	0.94	0.94

* Significant at 10%. ** Significant at 5%. *** Significant at 1%.

Provincial capital cities' SO_2 concentration did not significantly affect provincial goals (Table 4.2). But in most provinces, capital cities only occupy a fraction of the total land area and thus could not represent the general picture. Another problem with this variable was its coefficient's sign. Intuitively, the sign should be negative – dirtier air needs more reduction of pollutant emissions. The actual coefficient, although not significant, was consistently positive (Models 1–3 in Table 4.2). To avoid its impact, the variable was excluded from other models. Second, provinces with higher emissions per capita did not face deeper reductions. Emissions per capita did not have any significant relationship with provincial reduction goals. Third, earlier efforts on SO_2 emission control were not awarded later with relaxed goals. Neither of the two relevant variables – SO_2 removal rates in 2005 and goal attainment in the 10th Five-Year Plan – showed any consistent significance. Earlier efforts did not make the future easier in SO_2 emission control, while no failure in the past would get punished through adding future burden. Because of the very high correlation between the goal attainment variable and provincial growth rates of SO_2 emissions in the 10th Five-Year Plan, the model results also indicated that faster emission growth did not have a significant impact on provincial goals. For China's political reality, this result was reasonable. Provincial and other local leaders often rotate every five years. If one administration was irresponsible, its failure did not get the next administration punished. Similarly, a performing administration should not reduce pressure on future leaders. Fourth, more electricity net export consistently led to less stringent goals, but the relationship was not statistically significant. It seemed that China did not take serious consideration of the disintegration between emissions and consumption in distributing environmental goals. Fifth, no influence was found solely due to the location of a province. Regional characteristics should have been absorbed into other variables. For example, long-term goals already considered prevalent wind and more damage from western SO_2 emissions. Western and central provinces were poorer than eastern ones, which was reflected in GDP per capita.

Three principles were distinguished for distributing the national SO_2 emission goal in the 11th Five-Year Plan: those provinces with heavier pollution, bigger total emissions and richer GDP per capita should reduce more. The second principle was the most consistently applied. An explicit formula of deciding a provincial goal could be written as

Provincial Goal (−0 to −100) = −1.34 × Nonpower emission density (tons/km^2) − 0.047 × Total emissions (10,000 tons) − 1.27 × GDP per capita (10,000 RMB/person).

The 27 provinces had an arithmetic average goal in the 11th Five-Year Plan of −10.1%. The formula would lead to −10.2%: GDP per capita, −2.0%; nonpower SO_2 emissions density, −4.0%; and total emissions, −4.3%. The explanatory power was high, with adjusted R^2 generally over 0.93 (Model 7 in Table 4.2).

3.2 Goal distribution from provincial to municipality governments

The SEPA issued guidance for distributing SO_2 emission goals from one government level to its subordinate level (SEPA, 2006a). The total emissions are distinguished into the power sector (capacity no less than 6 MW) and nonpower sectors (SEPA, 2006a). The SO_2 emission quota was generally assigned to each fossil-fuel power plant according to provincially homogeneous emission intensity (grams SO_2/kWh, varying with plant ages; SEPA, 2006a). As shown in Figure 4.1, the designated emission intensity was more stringent in new coal power plants and those in eastern or richer provinces. From provinces to municipalities, polluting sources in nonpower sectors received their upper limits on the basis of achieving local air quality – particularly SO_2 emissions concentration with a threshold of 0.060 mg/m³ (SEPA, 2006a). The guidance did not clarify everything for assigning goals. It left decisions to provincial governments, especially in nonpower sectors. More important, the excess emission quota of a region was allowed to transfer or trade across regions (SEPA, 2006a).

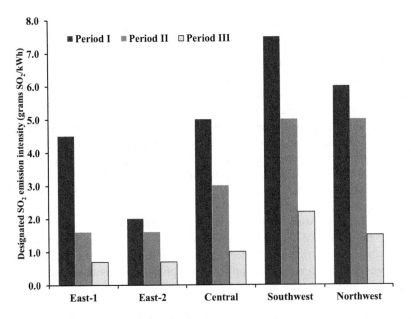

Figure 4.1 Designated SO_2 emission intensity in distributing SO_2 emissions quota to coal-fired power plants for 2010 in the 11th Five-Year Plan

Source: SEPA (2006a).

Note: Coal-fired power plants falling in Period I refer to those that went online or passed the Environmental Impact Assessment reports before December 31, 1996. Period II spans from January 1, 1997, to December 31, 2003. And Period III is from January 1, 2004, to the present. "East-1" includes the provinces of Liaoning, Hebei, Shandong, Zhejiang, Fujian and Hainan. "East-2" covers Beijing, Tianjin, Shanghai and Jiangsu Provinces. "Central" refers to Heilongjiang, Jilin, Shanxi, Henan, Hubei, Hunan, Anhui and Jiangxi Provinces. "Southwest" provinces are Chongqing, Sichuan, Guizhou, Yunnan, Guangxi and Tibet. "Northwest" has Inner Mongolia, Shaanxi, Gansu, Ningxia, Qinghai and Xinjiang.

In distributing provincial goals, official documents often did not even qualitatively declare what factors took effect. Furthermore, municipality data were not as publicly available as provincial data. Two aspects receive special attention in examining the provincial scheme of goal distribution: (1) whether the same principles in the national goal distribution held and (2) whether the SEPA's guidance was followed. This section looks at four provinces: Hebei, Guangdong, Jiangsu and Shanxi. Statistical model results are given in Table 4.3. Those models without significance are not shown. According to how the four provinces obey the national principle – rich and big provinces reduce more – a matrix is generated in Table 4.4.

Provinces have high autonomy in further allocating their goals among municipalities. The four provinces are distinguished with different patterns to demonstrate such decentralized authority. In Hebei Province, all municipalities got roughly same reduction goals. The provincial goal of Hebei Province was a 15% reduction from 1.50 million tons in 2005 (State Council, 2006). It had 11 municipalities, and their SO_2 emissions in 2005 ranged from 45,000 to 311,000 tons (Hebei Provincial Government, 2007). GDP per capita also varied, from 9,900 to 27,900 RMB/person (Hebei Provincial Statistics Bureau, 2006). The 2005 data on the power sector's share in SO_2 emissions are not publicly available. Information from the goals for 2010 is applied instead: the share of the power sector would range from 13.4% to 53.4% in the plan (Hebei Provincial Government, 2007). However, municipality goals only varied from a 14.1% to a 15.8% reduction, centering on the provincial goal (Hebei Provincial Government, 2007). Statistical models indicated a consistently significant constant (Table 4.3). But neither GDP per capita nor SO_2 emissions showed any impact on municipality goals. If the goal distribution guidance from the central government worked, a municipality with more SO_2 emissions from the power sector should receive somewhat more stringent goals. But no negotiation seemed to have shaped the municipality goals. The provincial government, very likely its top leaders, decided that the provincial goal was applied to all with minor adjustments.

In Guangdong Province, higher income and more SO_2 emissions led to more stringent goals. In the 11th Five-Year Plan, Guangdong Province received a

Table 4.3 Regression model results for distributing provincial goals to municipalities

	Hebei	*Guangdong*	*Jiangsu*	*Shanxi*
Observations	11	15	13	11
Nonpower emission density in 2005			−1.5**	
GDP/capita in 2005		−7.2***	6.2***	
SO_2 emissions in 2005		−3.3***	−1.4**	−0.98***
Power sector's emission share in 2005			−58.1***	
_Constant	−15.1***	33.9***	21.4**	
Adjusted R^2	0.90	0.76	0.86	0.87

Note: Six municipalities in Guangdong province with 2005 SO_2 emissions no more than 11,000 tons are not included in the model.
* Significant at 10%. ** Significant at 5%. *** Significant at 1%.

provincial goal of a 15% reduction from 1.29 million tons in 2005 (State Council, 2006). Among its 21 municipalities, 5 emitted less than 10,000 tons in 2005, and another 13, no more than 60,000 tons (Guangdong Environmental Protection Bureau, 2006). The biggest three emitted 475,000 tons, taking 51% of the provincial emissions that belonged to municipalities (929,000 tons; the remaining 365,000 tons were directly claimed to the provincial level; Guangdong Environmental Protection Bureau, 2006). Collectively, these three should reduce their SO_2 emissions by 46% (Guangdong Environmental Protection Bureau, 2006). The other 18 municipalities were even allowed to increase their emissions by 23% (Guangdong Environmental Protection Bureau, 2006). Regression models could better distinguish influential factors. To avoid the heavy impacts of outlying data points, six municipalities were excluded, with total SO_2 emissions in 2005 being no more than 11,000 tons and their goals in 2010 allowed for over 170% growth. Models for the rest of the 15 municipalities show significance of GDP per capita and SO_2 emissions (Table 4.3). Nonpower SO_2 emissions density was not tested because of data unavailability. Different from the situation for provincial goals, the constant variable here is significant. For every 10,000 RMB/person increase of GDP per capita and 10,000 tons more of SO_2 emissions in 2005, a municipality goal would be, respectively, 7.2% and 3.3% more stringent (from the 2005 level). Although the coefficients were different from those for provincial goals, the qualitative principles remained the same: rich and big municipalities should reduce more.

In Jiangsu Province, municipalities with higher emissions should reduce more, but richer ones were allowed to reduce less. Jiangsu Province's goal was an 18% reduction from 1.23 million tons in 2005 (State Council, 2006). The 13 municipalities emitted from 28,000 to 243,000 tons of SO_2, a much narrower but still larger range than in Guangdong Province (Jiangsu Provincial Government, 2008). SO_2 emission goals varied between 2.6% to 53.6% reduction (Jiangsu Provincial Government, 2008). Regression models included four independent variables, all for 2005 at municipality level: nonpower SO_2 emissions density, GDP per capita, SO_2 emissions and the power sector's share in total emissions (Table 4.3). The constant variable showed significance, and its appearance in models made the adjusted R^2 bigger. Corresponding to the increase of, respectively, 10,000 tons of SO_2 emissions, 1% of the power sector's share and 1 ton/km^2 of nonpower SO_2 emissions density, a municipality goal would become 1.4%, 0.6% and 1.5% more stringent (from the 2005 level). The signs of these coefficients were all reasonable and consistent with the situation of distributing the national goal to provinces, but GDP per capita displayed the opposite effect: for a municipality with 10,000 RMB/person richer, its SO_2 emissions were allowed to grow by 6.2%. For a province, this strategy of "rich municipalities reduce less" might maximize its GDP as well as tax income through entitling more opportunities to more promising municipalities.

In Shanxi Province, municipalities with higher emissions should reduce more, but income level did not have significant impacts. Shanxi Province's goal was a 14% reduction in the 11th Five-Year Plan from 1.52 million tons in 2005 (State

Table 4.4 Provincial goal distribution matrix

		Rich guys reduce		
		Less	*Neutral*	*More*
Big guys reduce	Less			
	Neutral		Hebei	
	More	Jiangsu	Shanxi	Guangdong

Council, 2006). Its 11 municipalities emitted from 92,000 to 185,000 tons of SO_2 in 2005, a much narrower range compared with the earlier three provinces or the national situation (Shanxi Provincial Government, 2006). Their GDP per capita in 2005 was between 5,500 and 26,000 RMB/person (Shanxi Bureau of Statistics, 2006). The municipality goals were scattered from a 8.5% to a 17.8% reduction (Shanxi Provincial Government, 2006). Only total SO_2 emissions showed a significant influence in the statistical models (Table 4.3). For emitting every 10,000 tons more of SO_2, a municipality goal would be about 1% more stringent (from the 2005 level). The GDP per capita's coefficient was negative, although not statistically significant. The principle – big provinces should reduce more – held here. That rich provinces should reduce more was not well applied but could have been considered.

In conclusion, provinces differed from each other in adopting principles from distributing the national goal (Table 4.4), which closely reflected that China's governance and, especially, environmental governance had been greatly decentralized (see Chapter 3). The most consistent principle across provinces was that bigger emitters should reduce more. The guidance from the SEPA did not have to be exactly followed.

4 Decentralized goal attainment

To keep the goal process running, goal attainment assessment is an inalienable step. It examines the effectiveness of the goal process and provides feedback. The key questions are what can be called goal attainment and how to evaluate it. For one Five-Year Plan, goal attainment evaluation does not wait until its conclusion. In the 11th Five-Year Plan, China publicized provincial SO_2 emissions every half a year (State Council, 2007c). At the end of 2008, a halfway assessment was scheduled (State Council, 2007c).

4.1 Criteria for goal attainment

In the 11th Five-Year Plan, China established "three systems" to facilitate SO_2 mitigation, which covered statistics, monitoring and evaluation (State Council, 2007b). This capacity building was planned and carried out by the SEPA and endorsed by the State Council (State Council, 2007b). Because of China's decentralization, that the central government mainly governs through provincial governments, the national system only targeted the provincial level (State Council, 2007b). The subordinate governments within provinces were evaluated with the rules passed along by provincial governments. For example, Zhejiang Province later enacted a more detailed, although not systematically different, regulation targeting municipality and county governments (Zhejiang Provincial Government, 2008).

Provincial goal attainment was evaluated with three criteria in the 11th Five-Year Plan (State Council, 2007b). The first criterion was on the quantitative goal itself and environmental quality. It was often a binary judgment: if they were attained, that was a mission accomplished. Excessive reduction would not be further awarded after the goal had been attained, while more emissions would not be punished if the goal was already broken. The second criterion was on the establishment and operation of three institutions: environmental goal setting of major pollutants, monitoring and goal attainment evaluation (State Council, 2007b). They were mainly judged by the enactment and distribution of official documents. The third one was on mitigation measures, including the completion and operation of pollutant removal facilities, the closure of inefficient factories, policy enactment and plan implementation (State Council, 2007b). If any of the three criteria failed to pass evaluation, the overall goal attainment would be judged a failure (State Council, 2007b). Accordingly, one feature was that the attainment of the goal itself, despite its central importance, did not ensure overall success. The first criterion focused on the results and the other two on the process. The regulation of the Chinese central government on the process provided feedback for local governments for adjusting policies and monitoring their implementation.

The 9th Five-Year Plan barely had any defined scheme to evaluate SO_2 goal attainment (NEPA et al., 1996). The official document available in the public domain only pointed out that the SO_2 control would be annually examined and evaluated and the result would be publicized periodically (NEPA et al., 1996). The SO_2 goal and its attainment process were better defined in the 10th Five-Year Plan, but no evaluation scheme was clearly defined either (SEPA, 2001). The National 10th Five-Year Plan for Environmental Protection only expressed several principles, including holding local government leaders responsible and linking environmental goal attainment with the leaders' performance evaluation (SEPA, 2001). The evolution path displayed China's progress in establishing a working evaluation scheme of SO_2 emission goals. Although still not perfect, the much clearer scheme in the 11th Five-Year Plan could have significantly contributed to the SO_2 mitigation goal attainment.

Recognizing the importance of credible data collection for achieving SO_2 emission goals in the 11th Five-Year Plan, China experienced an intensive capacity-building process. In December 2006, the updated *Management Methods of Environmental Statistics* entered into force (SEPA, 2006c). The regulation specified the organization and personnel for environmental statistics, rules on environmental survey and management and publication of environmental data (SEPA, 2006c). For goal attainment in the 11th Five-Year Plan, China strengthened its statistical system focusing on data credibility. SO_2 emissions were divided into three categories: power, nonpower industries and domestic (State Council, 2007b). The first two categories (industrial sectors) were further distinguished into two – key and non-key surveyed sources – based on the sizes of emission sources, and key surveyed sources covered 65% of total industrial SO_2 emissions (State Council, 2007b). Three parallel methods were applied under various situations: direct monitoring, estimation according to sulfur budget and estimation according to emission factors (State Council, 2007b). The first method, if applicable, had the highest priority (State Council, 2007b). SO_2 emissions from non-key surveyed sources were estimated following a similar trend as key surveyed sources (State Council, 2007b). Data about coal consumption and sulfur contents worked out SO_2 emissions from domestic sectors (State Council, 2007b). In addition, if cheating were caught in an SO_2 removal facility more than twice a year, no SO_2 removal would be recognized in the statistics data from the facility (State Council, 2007b). Furthermore, another two significantly more detailed policies were enacted for building emission inventories (SEPA, 2007a, 2007d). Not only were detailed accounting methods clearly written, but also the data report was regulated in specifics (SEPA, 2007a, 2007d).

4.2 Incentives for goal attainment

The central government has decentralized its power greatly since the economic reform started in 1978. As discussed in Chapter 3, three measures could exist to incentivize the cooperation of local governments by targeting local leaders, administrative constraints and fiscal transfer. Corresponding to the personnel relationship that is mainly established across the various levels of the Chinese Communist Party, the top national leadership of the party can greatly decide the promotion and removal of provincial-level leaders. The attainment of key goals, including those on environmental protection and SO_2 mitigation, had become an important aspect in the evaluation of provincial leaders' job performance. Concerning SO_2 mitigation goal implementation since the 11th Five-Year Plan, local government leaders but not local environmental protection bureau (EPB) leaders were targeted. The clear evidence was that provincial deputy governors, not EPB directors, were required to sign pollutant emission control contracts with the central government (SEPA, 2006b). The failure in the 10th Five-Year Plan on surging coal consumption demonstrated that pollution control had been far beyond the responsibility of EPBs alone.

Officially five characteristics distinguish a leader in the Chinese Communist Party for promotion or removal: virtue, ability, diligence, achievements and absence of corruption (The Central Committee of the Chinese Communist Party, 2002). Furthermore, after the formation of "Scientific View of Development," resource consumption, environmental protection and sustainable development were clearly pointed out to comprise "achievements" (Department of Organization of the Chinese Communist Party, 2006). Contracts on pollutant emission control and energy conservation clarified even more the responsibilities of local government leaders (SEPA, 2006b). Two institutions were applied for the attainment of the SO_2 emission goal in the 11th Five-Year Plan: accountability and veto (State Council, 2007a). "Accountability" demanded local government leaders be held accountable for their governance that fell within their jurisdictions. For example, the administrator of the SEPA, Xie Zhenghua, was forced to resign in 2005 for a serious pollution event in the Songhua River. "Veto" meant that local government leaders would fail evaluation on their entire job performance if the SO_2 emission goal were not attained. Promotion became inappropriate for these leaders. If goal failure did not degrade the leaders' ranks, they may still face a risk of being removed from original positions to some less significant ones. On the other hand, successful goal attainment was an important achievement and could help the leaders' promotion. A recently developed method for targeting local leaders has been gradually promoted by the Ministry of Environmental Protection (MEP) and later by the Ministry of Ecology and Environment. Top leaders of those provinces and municipalities that show serious environmental problems or fail environmental goals are forced to have "interview appointments" with the ministry (Ministry of Ecology and Environment, 2020a). Although those local leaders may not face immediate consequences of punishment, they will receive crucial warnings that darken their future promotion opportunities, especially if no quick fix is achieved afterward.

Another mechanism that was applied in the 11th Five-Year Plan was to temporarily constrain local administrative authorities as punishment: if a goal was not attained, no new construction projects would receive the ratification of their environmental impact assessment (EIA) reports for a given period. Over the 11th Five-Year Plan period, large construction projects still demanded ratification from the central government. In terms of environmental protection, every project with potential environmental damage should compose an EIA report and submit for ratification to various levels of governments (National People's Congress, 2002). The SEPA, and later the MEP, at the central level was responsible for large projects, such as new coal-fired power plants over 200 MW (SEPA, 2002). No project without the MEP's ratification could legally start construction. In early 2007, the SEPA temporarily suspended ratifying EIA reports of four municipalities and four power corporations (SEPA, 2007c). The suspension took effect for three months to force their cooperation (SEPA, 2007b). Afterward, the policy was formally established to target goal failure (SEPA, 2008). A failure to achieve the SO_2 emission goal could result in regional suspension for one month, three months or half a year. If no satisfying progress were made, the suspension could even last longer until full cooperation. The "suspension" policy may seriously influence the regional economy. Since GDP is the most important criterion in evaluating local leaders, this mechanism could effectively force cooperation. Capital investment was a crucial part of China's GDP. For example, in 2007, China's overall GDP was 24.7 trillion RMB, and capital investment comprised 13.7 trillion RMB, about 56% (National Statistics Bureau, 2008). A one-month suspension could delay construction and significantly affect capital investment and, consequently, the local economy. GDP growth itself occupied the most important status in evaluating local government leaders. In addition, a booming GDP could provide growing tax income not only to make officials more powerful but also to enable more budgets for poverty alleviation, health care, education and other key governmental affairs. Many of these issues are closely connected with the evaluation of leaders.

Fiscal transfer has not been explicitly linked with environmental goal attainment. However, the very significant fiscal transfer from the central to local governments (as discussed in Chapter 3), if institutionally associated with pollutant emission control, is potentially powerful to mobilize local governments for environmental protection.

With China's further decentralization of governmental authorities, the first mechanism to directly target local governments is expected to be even more important. In the past four decades, the central government has been continuously loosening direct management of local governmental affairs. As a key feature of the economic reform, China has greatly reduced the requirements of administrative ratification and decentralized much remaining authority to local governments (State Council, 2013b, 2014). Fossil-fuel-fired power plants were no longer required for the MEP's ratification after 2015 and the authority entirely went to provincial governments (MEP, 2015; Ministry of Ecology and Environment, 2019).

5 Goal evolution

Corresponding to different strategies for controlling air pollution–induced health damages, three major types of goals can be adopted. First, emission mitigation goals of key pollutants, prominently SO_2, aim to directly target the sources of environmental pollution. The second type focuses on controlling air pollutant concentrations. Ambient air quality standards are widely adopted across countries to specify concentration thresholds of key air pollutants individually, such as SO_2, fine particulate matter ($PM_{2.5}$) and ozone (O_3). As discussed earlier, the control of ambient SO_2 concentration was a key scientific foundation to decide China's long-term SO_2 mitigation goal at 12 million tons (Yang et al., 1999). The third type targets environmental quality directly through the Air Quality Index (AQI) that provides a synthesized measurement of key air pollutant concentrations. The AQI also guides people's activities corresponding to air quality conditions. Although the three strategies have a similar ultimate goal for protecting public health, they have different implications for implementation. Local governments can only directly mitigate local emissions while local pollutant concentration is determined by emissions within and outside of their jurisdiction as well as weather conditions, land use and other factors. Then their motivation could differ significantly under the different types of goals to affect their performance of pollution mitigation.

Over the past two decades, China has been switching back and forth between major governance strategies on environmental protection with different types of goals. As clearly stated in China's environmental protection law, local governments are responsible for environmental quality within their jurisdictions (National People's Congress, 1989). However, environmental protection was not ranked high among all governmental tasks in the 1990s. Local leaders generally prioritized economic growth for promotion opportunities. The 10th Five-Year Plan (2001–2005) was a transitional period toward the Total Emission Control regime to set up environmental goals for reducing major pollutant emissions by 10% (National People's Congress, 2001). However, due to the lack of environmental cleanup incentives and the acceleration of economic growth, SO_2 emissions went up by 27.8%, and only 2 out of 31 provinces achieved their allocated goals. Demand for serious, effective and efficient compliance monitoring had not been strong. The 11th Five-Year Plan (2006–2010) was a milestone in China's environmental protection history. The Total Emission Control regime was strengthened, while serious and implementable incentives were put into place for local governments to achieve their individual mitigation goals (Xu, 2011). A bottom-up compliance monitoring system on emissions was initiated and established (SEPA, 2007d). Although SO_2 emissions did decline in the 11th Five-Year Plan, data manipulation also strained the compliance monitoring system as indicated in the gaps between official and independent emission inventories (Lu et al., 2011).

Concerning SO_2 emissions, two sets of regulations were most important and direct, being effluent emission standards and ambient air quality standards. Previously, cities were given goals of "blue sky" days. "Blue sky" was defined as that air quality reached the Grade 2 standard. One crucial change in the 2012 version ambient air quality standards was the addition of $PM_{2.5}$ (MEP, 2012; National Environmental Protection Administration and State Bureau of Technical Supervision, 1996). $PM_{2.5}$ concentration is more closely related to air quality that affects public health, while the emissions of SO_2 and other pollutants are only indirect measures. In other words, $PM_{2.5}$ goals are more related to ends of air pollution control, while SO_2 emissions goals are more about means. $PM_{2.5}$ comprises many more pollutants, including sulfate particles that are originated from SO_2 emissions.

Together with the 2012 update of the ambient air quality standards, China enacted the Ambient Air Quality Index (Ministry of Environmental Protection, 2012). It synthesizes key air pollutant concentrations into one index to indicate air quality. The cutoff AQIs between "excellent," "good" and "polluted" air are 50 and 100, respectively. Each air pollutant can calculate its individual AQI (IAQI) and the composite AQI is the largest IAQI, or the IAQI of the primary air pollutant. An AQI of 50 or lower corresponds to the Grade 1 ambient air quality standards, while 100 or lower corresponds to Grade 2. They provide the technical foundation for China to adopt regulatory strategies that are based on air quality rather than pollutant emissions. Both regulations gave nearly four years of grace periods and formally entered into force in January 2016. The 12th Five-Year Plan (2011–2015) initially continued with the Total Emission Control scheme to include more pollutants (National People's Congress, 2011). However, a major air pollution episode in January 2013 that badly hit North China, most notably Beijing, pushed the Chinese government to rethink its strategy (State Council, 2013a) and accelerated the shift toward the air quality approach and the application of the two related standards.

Air quality goals and emission mitigation goals of SO_2, as well as other major air pollutants, both aim for public health benefits. In order to achieve air quality goals that focus on ambient air pollution, efforts should still primarily fall on the mitigation of pollutant emissions together with their geographic and temporal distributions. Due to the atmospheric transport of air pollution, the attainment of $PM_{2.5}$ goals depends not only on a region's own mitigation efforts but also that of neighboring regions. The interregional reliance tends to be greater for geographically smaller jurisdictions. Accordingly, free riding may be a potential problem to compromise the willingness to engage in hard mitigation efforts. Nevertheless, data credibility is a key element in enforcing environmental policies as well as the top-down goals. Emission mitigation data, however, tend to be much more conveniently manipulated than air quality data. The number of polluting sources in China could easily overwhelm its compliance monitoring resources, especially in sparsely populated and less developed regions. In the 11th Five-Year Plan, the MEP assembled teams to inspect provinces and their polluting firms. However, the data had been of unsatisfying quality, and what was reported by local governments and polluting firms was often seriously discounted. Data on SO_2 emissions are more prone to manipulation because the bottom-up monitoring and reporting have to go through many stakeholders who have incentives to underreport emissions and overreport mitigation. Occasional verification from the central government often finds big gaps in data and must "squeeze moisture" from the reported mitigation amounts. In contrast, ambient air quality data are much more difficult to manipulate and any dishonest behavior is much easier to discover. The central government also runs its own air quality monitoring network via ground stations and remote sensing, such as satellites. Accordingly, China reversed the strategy to have air quality improvement targets (State Council, 2013a). Air quality monitoring stations are much fewer than polluting sources to substantially reduce the resource burden of compliance monitoring. Thus, the probability of compliance, together with the better data quality, should be much higher.

The prospective penalty and reward for goal attainment do not differ substantially from the 11th Five-Year Plan to the 12th and 13th. However, the 12th and 13th Five-Year Plans achieved much faster SO_2 mitigation, even considering the slower economic growth rates. It could indicate that the free-riding problems were less important than data credibility. Furthermore, SO_2 emissions are just one among many pollutants, while $PM_{2.5}$ could better serve as a comprehensive air quality indicator. Provincial and local governments could have greater flexibility in weighing various technological and policy mitigation alternatives. It could also potentially encourage more local policy innovations and probably achieve better cost-effectiveness through balancing the marginal abatement costs of pollutants.

Furthermore, although emission reduction goals have been consistently achieved since the 11th Five-Year Plan, air quality was not perceived to have improved. One possible cause could be the problems in reporting emission data,

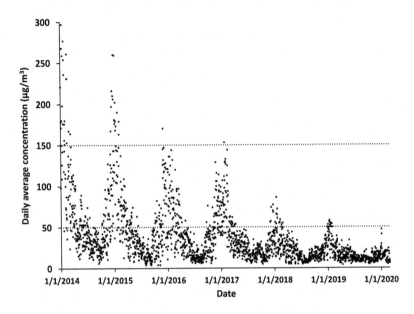

Figure 4.2 Daily SO₂ concentrations in Shijiazhuang (1 January 2014–29 February 2020)

Source: Ministry of Ecology and Environment (2020b).

Note: The upper and lower dotted horizontal lines indicate the Grade 2 and 1 standard, respectively, in China's ambient air quality standards in 1996 and 2012.

as discussed earlier. Another more important reason for the wide gap between the successful attainment of SO₂ mitigation goals and the perceived terrible air quality was that SO₂ has been increasingly less important in ambient air quality. For example, Hebei Province often has one of the highest anthropogenic PM₂.₅ concentrations in China and the world. In its capital city, Shijiazhuang, air quality is taken as one example to illustrate the importance of new PM₂.₅ standards and goals. Significant improvements have been made on reducing SO₂ emissions and concentrations. In the first three months of 2014, SO₂ concentrations in Shijiazhuang exceeded the Grade 1 standard (50 μg/m³) in 90% of all days, while a strong seasonal cycle indicated that the winter or the heating season as the worst season (Figure 4.2). From February 2019 to February 2020, in contrast, the standard was not exceeded for even a single day (Figure 4.2). It illustrates China's hard and effective efforts in controlling SO₂ emissions and bringing down SO₂ concentrations. Essentially the original long-term goal for SO₂ mitigation, 0.060 mg/m³ or 60 μg/m³ (SEPA, 2006a), had been generally achieved. However, from the perspective of PM₂.₅, Shijiazhuang's performance has been much less impressive. Its concentration has regularly exceeded the much more relaxed Grade 2 standard (Figure 4.3).

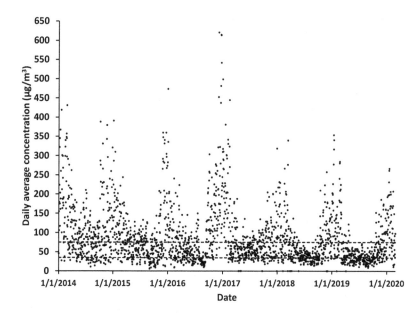

Figure 4.3 Daily PM$_{2.5}$ concentrations in Shijiazhuang (1 January 2014–29 February 2020)

Source: Ministry of Ecology and Environment (2020b).

Note: The upper and lower dotted horizontal lines indicate Grade 2 and 1 standard, respectively, in China's 2012 ambient air quality standards.

PM$_{2.5}$ is not a single pollutant but a set of various pollutants that fall into the size range. SO$_2$ is a gaseous pollutant and could be converted into sulfate particles in the atmosphere to become one important component of PM$_{2.5}$. Heating seasons in northern China tend to result in more coal consumption and pollutant emissions, while inversion (when warm air is above cold air) is more frequent in the winter when the ground is cold, suppressing convection, and thus facilitates the accumulation of pollutant concentrations. Although SO$_2$ is one key precursor species of PM$_{2.5}$, other air pollutants are also crucial components in forming PM$_{2.5}$.

Furthermore, ozone pollution has significantly deteriorated over the period. O$_3$ and PM$_{2.5}$ concentrations tend to have opposite seasonal cycles. Chemical reactions to form O$_3$ in the atmosphere involve nitrogen oxides (NOx), volatile organic compounds (VOC) and sunlight, while summer months tend to provide more favorable conditions. PM$_{2.5}$ and SO$_2$ concentrations peak in winter months, and O$_3$–8h concentration (daily maximum concentration over 8 hours) is the highest in summer months (Figure 4.4). As a result, mitigation goals of SO$_2$ emissions and SO$_2$ concentrations will be at a greater distance from perceived air quality that mainly corresponds to PM$_{2.5}$ and O$_3$ concentrations.

SO$_2$ has never been the primary pollutant to decide Shijiazhuang's monthly AQI since 2014 (Figure 4.5). PM$_{2.5}$ dominated the AQI before 2016, while in and

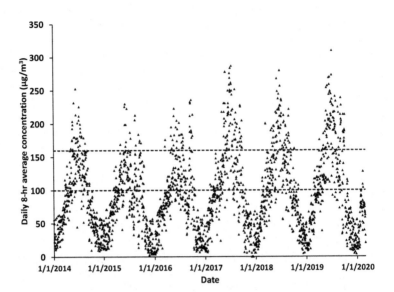

Figure 4.4 Daily 8-hour O₃ concentrations (daily maximum concentration over 8 hours) in Shijiazhuang (1 January 2014–29 February 2020)

Source: Ministry of Ecology and Environment (2020b).

Note: The upper and lower dotted horizontal lines indicate Grade 2 and 1 standard, respectively, in China's 2012 ambient air quality standards.

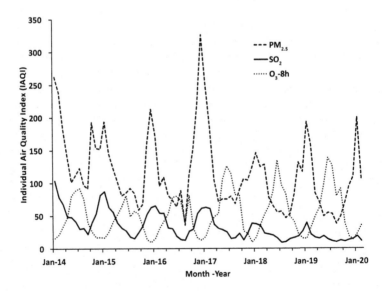

Figure 4.5 Monthly average AQI in Shijiazhuang (January 2014–February 2020; calculated from daily data)

Source: Ministry of Ecology and Environment (2020b).

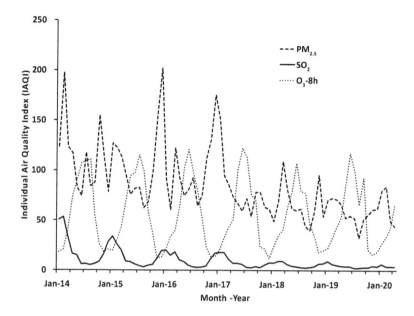

Figure 4.6 Monthly average AQI in Beijing (January 2014–February 2020; calculated from daily data)

Source: Ministry of Ecology and Environment (2020b).

after 2017 with reduced $PM_{2.5}$ concentration and rising O_3–8h concentration, O_3 became the primary pollution in summer months and $PM_{2.5}$ remained dominant in the winter (Figure 4.5). The trend is similar in other Chinese cities. For example, Beijing witnessed the rise of O_3 in determining summer AQI a few years earlier than Shijiazhuang did (Figure 4.6). In southern China, where winter is mild/warm with adequate sunshine, the importance of O_3 entirely overshadows that of $PM_{2.5}$ in the AQI. For example, in Shenzhen, AQI in most months is now decided by O_3–8h but not $PM_{2.5}$ (Figure 4.7).

From 2014 to 2020, $PM_{2.5}$ concentrations and corresponding air quality indexes have been reduced throughout major cities in China, but O_3–8h generally had a rising trend. One reason for their diverging trends in the past years could be traced to the presence of $PM_{2.5}$ goals but not O_3 goals. In the 13th Five-Year Plan, China further enacted air quality goals together with 15% reduction goals on SO_2 and NO_x emissions (National People's Congress, 2016). The proportion of days that the AQI is below 100 in municipalities should reach 80%, while for those cities with $PM_{2.5}$ concentrations not reaching the Grade 2 standard (or 75 µg/m³), they should reduce the level by 18% over the five years (National People's Congress, 2016).

The AQI is a more comprehensive measure of air pollution to consider both $PM_{2.5}$ and O_3. In China's further goal evolution especially into the 14th Five-Year

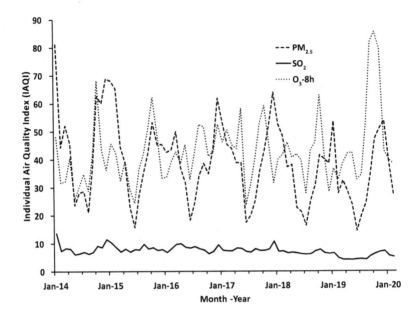

Figure 4.7 Monthly average AQI in Shenzhen (January 2014–February 2020; calculated from daily data)

Source: Ministry of Ecology and Environment (2020b).

Plan (2021–2025), it could play a more prominent role in mobilizing local governments for air pollution control.

Note

1   This chapter is based on the author's own material used in Xu, Y. 2011. The use of a goal for SO_2 mitigation planning and management in China's 11th five-year plan. *Journal of Environmental Planning and Management*, 54, 769–783; much of which has been revised and expanded on.

References

The Central Committee of the Chinese Communist Party. 2002. *Regulations on selecting and appointing leaders of the party and governments*. Beijing, China: The Central Committee of the Chinese Communist Party.

Chakravarty, S., Chikkatur, A., De Coninck, H., Pacala, S., Socolow, R. & Tavoni, M. 2009. Sharing global CO_2 emission reductions among one billion high emitters. *Proceedings of the National Academy of Sciences of the United States of America*, 106, 11884–11888.

Chinese Academy for Environmental Planning (CAEP). 2004. *Basic thoughts on national 11th five-year plan on environmental protection*. Beijing, China: CAEP.

Department of Organization of the Chinese Communist Party. 2006. *Temporary methods to evaluate local leaders for realizing scientific view of development*. Beijing, China: Chinese Communist Party.

DeYoung, K. 2009. Obama sets timetable for Iraq. *Washington Post*, February 28.

Guangdong Environmental Protection Bureau. 2006. *The 11th five-year plan on major pollutants emission goals of municipalities in Guangdong province*. Guangzhou, China: Guangdong Environmental Protection Bureau.

Hebei Provincial Government. 2007. *The 11th five-year plan on environmental protection of Hebei province*. Shijiazhuang, China: Hebei Provincial Government.

Hebei Provincial Statistics Bureau. 2006. *Hebei economy statistics yearbook*. Beijing, China: China Statistics Press.

Jiangsu Provincial Government. 2008. *The 11th five-year plan on environmental protection and ecological construction of Jiangsu province*. Nanjing, China: Jiangsu Provincial Government.

Li, J. 2010. *How to allocate CO_2 mitigation responsibilities within China – application of a new burden-sharing scheme*. Ph.D. dissertation, Princeton University Press, Princeton, NJ.

Liu, G., Zhang, Z., Dong, Z. & Wu, L. 2006. *Research report on China's ten five-year plans*. Beijing, China: People's Press.

Locke, E. A., Saari, L. M., Shaw, K. N. & Latham, G. P. 1981. Goal setting and task-performance – 1969–1980. *Psychological Bulletin*, 90, 125–152.

Lu, Z., Zhang, Q. & Streets, D. G. 2011. Sulfur dioxide and primary carbonaceous aerosol emissions in China and India, 1996–2010. *Atmospheric Chemistry and Physics*, 11, 9839–9864.

Ma, K. 2005. *Strengthen the scientific and democratic planning process and write a good 11th five-year plan*. Beijing, China: The 4th Conference of the 10th National People's Congress.

MEP. 2012. *Ambient air quality standards*. GB 3095-2012. Beijing, China: MEP.

MEP. 2015. *Catalogue of construction projects for environmental impact assessment ratification from the ministry of environmental protection (2015 version)*. Beijing, China: MEP.

Ministry of Ecology and Environment. 2019. *Catalogue of construction projects for environmental impact assessment ratification from the ministry of environmental protection (2019 version)*. Beijing, China: Ministry of Ecology and Environment.

Ministry of Ecology and Environment. 2020a. *Method for interview appointment by the ministry of ecology and environment (Draft for comments)*. Beijing, China: Ministry of Ecology and Environment.

Ministry of Ecology and Environment. 2020b. *Realtime city air quality data* [Online]. Available: www.mee.gov.cn/.

Ministry of Environmental Protection. 2008. *Assessment reports on provincial emissions of major pollutants*. Beijing, China: Ministry of Environmental Protection.

Ministry of Environmental Protection. 2012. *Technical regulation on ambient air quality index (on trial)*. Beijing, China: Ministry of Environmental Protection.

National Bureau of Statistics. 1997–2008. *China energy statistical yearbook*. Beijing, China: China Statistics Press.

National Bureau of Statistics. 2006. *China statistical yearbook*. Beijing, China: China Statistics Press.

National Bureau of Statistics. 2019. *China statistical yearbook*. Beijing, China: China Statistics Press.

National Bureau of Statistics of China. 1999. *China statistical yearbook*. Beijing, China: China Statistics Press.

National Environmental Protection Administration & State Bureau of Technical Supervision. 1996. *Ambient air quality standard.* GB 3095-1996. Beijing, China: NEPA.

National People's Congress. 1989. *Law of environmental protection.* Beijing, China: The 4th Conference of the 10th National People's Congress.

National People's Congress. 2001. *The outline of national 10th five-year plan on economic and social developments.* Beijing, China: The 4th Conference of the 9th National People's Congress.

National People's Congress. 2002. *Law of environmental impact assessment.* Beijing, China: The 4th Conference of the 10th National People's Congress.

National People's Congress. 2006. *The outline of the national 11th five-year plan on economic and social development.* Beijing, China: The 4th Conference of the 10th National People's Congress.

National People's Congress. 2011. *The outline of the national 12th five-year plan on economic and social development.* Beijing, China: The 4th Conference of the 10th National People's Congress.

National People's Congress. 2016. *The outline of the 13th five-year plan on economic and social development.* Beijing, China: The 4th Conference of the 10th National People's Congress.

National Statistics Bureau. 2008. *China's economic growth in 2007.* Beijing, China: China Statistics Press.

National Statistics Bureau & Ministry of Ecology and Environment. 2019. *China statistical yearbook on environment 2018.* Beijing, China: China Statistics Press.

NDRC. 2003. *Establish a new view of development and do well in preparing writing the 11th five-year plan.* Beijing, China: NDRC.

NDRC. 2005. *On the basic thoughts of the national 11th five-year plan.* Beijing, China: NDRC.

NEPA, National Planning Commission & NETC. 1996. *The national 9th five-year plan on controlling major pollutants' emissions.* Beijing, China: NEPA, NPC, NETC.

NEPA & SBTS. 1996. *Ambient air quality standard.* GB 3095-1996. Beijing, China: NEPA, SBTS.

Pacala, S. W. & Socolow, R. H. 2004. Stabilization wedges: Solving the climate problem for the next 50 years with current technologies. *Science,* 305, 968–972.

Pouyat, R. V. & McGlinch, M. A. 1998. A legislative solution to acid deposition. *Environmental Science & Policy,* 1, 249–259.

Roberts, L. 1991. Acid-rain program – mixed review. *Science,* 252, 371.

SEPA. 2001. *National 10th five-year plan on environmental protection.* Beijing, China: State Environmental Protection Administration.

SEPA. 2002. *Regulation on ratifying environmental impact assessment reports of construction projects at different government levels.* Decree No. 15. Beijing, China: State Environmental Protection Administration.

SEPA. 2005. *Plan on controlling acid rain and SO_2 emissions (draft for comments).* Beijing, China: State Environmental Protection Administration.

SEPA. 2006a. *Guidelines on calculating SO_2 emission quotas.* Beijing, China: State Environmental Protection Administration.

SEPA. 2006b. *A letter on signing up liability contracts of major pollutants.* Beijing, China: State Environmental Protection Administration.

SEPA. 2006c. *Management methods of environmental statistics.* Beijing, China: State Environmental Protection Administration.

SEPA. 2006d. *Monthly information on writing the national 11th five-year plan on environmental protection.* Beijing, China: State Environmental Protection Administration.

SEPA. 2007a. *Detailed methods to verify major pollutants emission reduction in the 11th five-year period (on trial)*. Beijing, China: State Environmental Protection Administration.

SEPA. 2007b. *Lvliang and Liupanshui municipalities and Huadian corporation are finally removed from suspension*. Beijing, China: State Environmental Protection Administration.

SEPA. 2007c. *SEPA disclosed illegal construction projects and used 'regional suspension of ratification' policy*. Beijing, China: State Environmental Protection Administration.

SEPA. 2007d. *Verification of major pollutants emission reduction in the 11th five-year period (on trial)*. Beijing, China: State Environmental Protection Administration.

SEPA. 2008. *Management methods of regional suspension of ratifying environmental impact assessment reports (on trial) (draft for comments)*. Beijing, China: State Environmental Protection Administration.

Shanxi Bureau of Statistics. 2006. *Shanxi statistics yearbook*. Taiyuan, China: Shanxi Bureau of Statistics.

Shanxi Provincial Government. 2006. *Notice on controlling major pollutants emissions in the 11th five-year plan*. Taiyuan, China: Shanxi Provincial Government.

State Council. 1990. *Regulations for the implementation of the law of standardization*. Beijing, China: State Council.

State Council. 2005a. *Decisions on realizing scientific view of development and strengthening environmental protection*. Beijing, China: State Council.

State Council. 2005b. *Several advices from the state council on strengthening the enactment of economic and social development plans*. Beijing, China: State Council.

State Council. 2006. *Opinion on pollution emission quota distribution in the 11th five-year plan*. Beijing, China: State Council.

State Council. 2007a. *Notice on distributing composite working plan on energy conservation and pollutant emission reduction*. Beijing, China: State Council.

State Council. 2007b. *Notice on distributing implementation plans and methods of statistics, monitoring and assessment on energy conservation and pollutant emission reduction*. Beijing, China: State Council.

State Council. 2007c. *Notice on distributing the 11th five-year plan on environmental protection*. Beijing, China: State Council.

State Council. 2013a. *Action plans on air pollution prevention and control*. Beijing, China: State Council.

State Council. 2013b. *Decisions on cancelling and decentralizing administrative ratification [2013(19)]*. Beijing, China: State Council.

State Council. 2014. *Decisions on cancelling and decentralizing administrative ratification [2014(5)]*. Beijing, China: State Council.

State Environmental Protection Administration (SEPA). 2001–2009. *National report on environmental statistics*. Beijing, China: State Environmental Protection Administration.

United Nations. 1992. *United nations framework convention on climate change*. New York: United Nations.

The U.S. Congress. 1990. *Clean air act amendments 1990*. Washington, DC: The U.S. Congress.

U.S. Environmental Protection Agency. 2007. *National emissions inventory (NEI) air pollutant emissions trends data*. Washington, DC: U.S. Environmental Protection Agency.

Wang, J., Wu, X., Cao, D. & Meng, F. 2004. Proposed scenarios for total emission control of SO2 during the 10th five-year plan period in China. *Research of Environmental Sciences*, 17, 4.

Wang, X. 2002. *Exponent on 'national 10th five-year plan on environmental protection'*. Beijing, China: Science Press.

Xinhua News Agency. 2005. *The birth of 'the suggestions to 11th five-year plan'* [Online]. Beijing, China [Online]. Available: http://news.xinhuanet.com/politics/2005-10/26/content_3685219.htm.

Xinhua News Agency. 2006. *The drafting of 11th five-year plan outline.* Beijing, China [Online]. Available: http://news.xinhuanet.com/politics/2006-03/16/content_4308918.htm.

Xu, Y. 2011. The use of a goal for SO_2 mitigation planning and management in China's 11th five-year plan. *Journal of Environmental Planning and Management*, 54, 769–783.

Yang, X., Gao, Q., Jiang, Z., Ren, Z., Chen, F., Chai, F. & Xue, Z. 1998. Research on the transportation and precipitation regular pattern of sulfur pollutants in China. *Research of Environmental Sciences*, 11, 27–34.

Yang, X., Gao, Q., Qu, J. & Jiang, Z. 1999. The exploration and initial assessment of total amount control method for SO_2 emission in China. *Research of Environmental Sciences*, 12, 17–20.

Zhejiang Provincial Government. 2008. *Notice on distributing Zhejiang provincial implementation plans and methods of statistics, monitoring and assessment on energy conservation and pollutant emission reduction.* Hangzhou, China: Zhejiang Provincial Government.

Zou, S., Wang, J. & Hong, Y. 2006. *Research report on national environmental protection plan in the 11th five-year plan.* Beijing, China: China Environmental Science Press.

Zou, S., Zhang, Z., Yan, G. & Tian, R. 2004. Preliminary assessment on medium-term execution of the national tenth five-year plan outline, important eco-environmental conservation and environmental protection plan. *In:* Wang, J., Zou, S. & Hong, Y. (eds.) *Environmental policy research series.* Beijing, China: China Environmental Science Press.

5   Policy making

1 China's challenges in policy making

Policies and goals are important in any country's governance, but their relative roles could have two primary patterns under different governance models. Rules are set up through policies (and laws), while polluters and other stakeholders decide on their own actions according to the rules. In the rule-based governance, policies are in the first place, while goals are more implicit to take the second place. Another strategy explicitly makes goals in the first place, while policies are secondary and could be more flexible. With the rule of law not yet well established, China would face great challenges in policy supply under the rule-based governance, especially given its rapidly evolving economy and society.

1.1 Uncertain linkages between actions and outcomes

China is rapidly industrializing and the economy grows at a fast pace. It encounters great uncertainties on whether planned actions could achieve intended goals. Sulfur dioxide (SO_2) emissions as well as other environmental problems tend to have a wide scope of influential economic, energy and environmental factors, as well as scattered emission sources in numerous important sectors. Many key factors for SO_2 mitigation are beyond the jurisdiction of environmental protection, specifically the Ministry of Ecology and Environment (and previously the Ministry of Environmental Protection). Implementation is largely under the responsibility of local governments, while the central government is not designed and well equipped for primary policy implementation. In addition, China's complexities can cast substantial uncertainties on whether preplanned actions can achieve their goals. China identified enough efforts to achieve the 10% reduction goals of SO_2 emissions in both the 10th and the 11th Five-Year Plans, but their outcomes differed from each other dramatically. In the *Outline* of the National 10th Five-Year Plan that was ratified by the National People's Congress, the 10% reduction goals of "major pollutants" were clearly written (National People's Congress, 2001). "Major Pollutants" were later defined to include SO_2, dust, COD (chemical oxygen demand), ammonia-nitrogen and industrial solid waste (SEPA, 2001). External measures, particularly energy conservation, did not show up in the national

Outline (National People's Congress, 2001). However, in the special plan for energy development, China did propose a goal to reduce energy intensity by about 15% to 17% and coal's share in total energy consumption by 3.88% in the five years (NDRC, 2001). China's annual economic growth rate, another key factor, was estimated to be 7% (National People's Congress, 2001). Between 2001 and 2005, the reversely calculated sulfur contents in coal from China's official data went down from 1.22% to 1.05% (Xu et al., 2009), and a lot more SO_2 scrubbers were installed (Figure 5.11). For the 10th Five-Year Plan, the SO_2 mitigation shortfall was mostly due to the unexpected surge in coal consumption as a result of accelerated economic growth, 87.6% over the five years that overwhelmed the efforts of the State Environmental Protection Agency (SEPA; BP, 2019).

In the 11th Five-Year Plan, the planning structure differed only slightly. With the same 10% reduction goal, the *Outline* of the National 11th Five-Year Plan narrowed the definition of "major pollutants" to cover only SO_2 and COD for primary attention (National People's Congress, 2006). Other pollutants were addressed in the special plan for environmental protection (State Council, 2007b). A goal on energy conservation, a 20% reduction of energy intensity, got promoted to the national *Outline*. Coal's share in total energy consumption remained in the special plan for energy development, with a 3% drop in the five years (NDRC, 2007). China's economy was estimated, or conservatively planned, to grow 7.5% per year (National People's Congress, 2006). These figures were quite close to those in the 10th Five-Year Plan. Simply from the planning perspective, these two 10% reduction goals of SO_2 emissions should both be attained. However, their results diverged significantly away from each other, which illustrated the difficulty to foresee the effects of policies and actions on goals.

1.2 Challenges in policy making to induce actions

In the U.S.'s efforts to control SO_2 emissions, the Clean Air Act Amendments (1990) established the Acid Rain Program that was distinguished as the most important law on the issue (The U.S. Congress, 1990). However, no individual environmental policy in China could claim an equal share of importance in its SO_2 mitigation cause. In comparison to goal-centered governance, policy supply under rule-based governance features fewer policies (or laws), and some are of crucial importance in achieving the intended goals of environmental protection. Each policy has a more extended enactment procedure and implementation horizon, which makes policy-making process lengthy and careful. The failure/success of a key policy thus takes on much heavier weight in environmental protection outcomes. Nevertheless, in developed countries where the rule of law is well established, the linkages between policies and polluters' actions are more predictable, while these actions further contribute to intended outcomes. However, because China has not established a sound rule of law, confidence is much lower that a policy can be implemented well to induce the intended actions.

Many environmental policy instruments have been designed and applied across countries. The first major category involves command and control policies, such as mandatorily shutting down polluting sources, setting pollutant emission and energy efficiency standards and mandating the application of the best available technologies. Another major category is based on economic incentives and markets. Typical policy instruments include effluent emission discharge fees, taxes, tradable permits and subsidies. Information disclosure, such as labeling and certificates, aims to enable consumers to voluntarily make informed consumption choices for minimizing environmental impacts.

Despite some unique features, China's policy toolbox for SO_2 mitigation was not fundamentally different from that in developed countries with rule-based governance. In China, an engineering approach that was based on SO_2 scrubbers in coal-fired power plants involved many command and control policies for their deployment and normal operation to meet effluent emission standards (Ministry of Environmental Protection and General Administration of Quality Supervision Inspection and Quarantine, 2011; the SEPA and General Administration of Quality Supervision Inspection and Quarantine, 2003). China has also been experimenting with market-incentive policies, such as cap-and-trade, an effluent emission fee or an emission tax (Yan et al., 2009; Dong et al., 2011; Ge et al., 2011; Zhang et al., 2016). Technological licensing from developed countries, through a functioning technology market, was a cornerstone in China's SO_2 mitigation to build a domestic industry for rapid deployment and cost reduction (Xu, 2011).

After assessing the effectiveness and efficiency of individual environmental policy instruments, policies are enacted for tackling a given environmental problem (Barron and Ng, 1996; Goulder and Parry, 2008). A few criteria could be important in making an optimal policy, including cost-effectiveness; capability to address uncertainty, synergy or conflict with current policy instruments; compliance monitoring and inspection capacity and requirements; and compliance of polluters. The latter two are especially relevant to developing countries like China, where the rule of law has not been well established and environmental noncompliance might be prevalent. In developed countries, there has been an increasing trend in the application of market-based instruments (Portney and Stavins, 2000; Tietenberg, 1990). Cost-effectiveness is the most important argument for their adoption considering particularly the reduced abatement costs (Goulder and Parry, 2008). For example, in the U.S. Acid Rain Program in the Clean Air Act Amendments (1990), total SO_2 emissions from coal power plants were capped and emission permits were allowed to trade in a market (The U.S. Congress, 1990). The policy substantially reduced the abatement costs compared with command-and-control instruments (Benkovic and Kruger, 2001).

Environmental policy instruments differ from each other in their capability of addressing uncertainties. For example, environmental taxes establish certain levels of emission prices but leave the quantities of emissions uncertain. In contrast, tradable permits with a fixed cap are more certain about the quantity within the defined boundary of emission sources but not about the price. Other instruments all have various impacts on uncertainties (Goulder and Parry, 2008). The introduction of a new environmental policy instrument should consider how it interacts with existing policies to create synergies or conflicts. If a new emission trading policy is imposed into an area that is already dominated by command and control policies, it may not be able to achieve its intended cost-effectiveness (Zhang et al., 2013). China's experiments of SO_2 emission trading schemes encountered major problems, including frequent governmental intervention and inter-policy conflicts together with the quality of policy design (Zhang et al., 2016).

Compared with developed countries, developing countries and specifically China have more difficulties in making optimal policies. Research tends to be thinner especially in the past to understand how individual policy instruments perform in their contexts. Significant constraints on environmental policy implementation may exist due to the lack of adequate financial resources, personnel and necessary expertise (Blackman, 2010). More details on China's policy implementation problems are discussed in Chapter 6. The effectiveness of individual policies could be very uncertain with unpredictable implementation, which makes policy design challenging.

2   Goal-centered policy supply

China's policy supply follows a very different pattern from that in rule-based governance. Goals play the central role in environmental governance, while policies as means to achieve goals are primarily instrumental and failures of individual policies are more accommodated. Important centralized goals as those few in National Five-Year Plans drive decentralized policies, laws and regulations from ministries, local governments, the People's Congress and other stakeholders. For those environmental fields without goals or with goals but at lower priorities, policy supply tends to be less adequate and strong.

2.1   Enabling goal-centered policy supply

Goal-centered governance in China is enabled by centralized national goals, decentralized goal attainment, decentralized policy making and implementation and mobilized central and local governments. In the past four decades, China's environmental governance has been heavily decentralized, as discussed in Chapter 3. The four levels of governments – central, provincial, municipality and county – have diverging divisions of governmental authorities and functions. As matched by their personnel categories and fiscal expenditures, the central government heavily focuses on policy making, while the county-level governments are almost entirely on policy implementation. Provincial- and municipality-level governments have significant authorities and functions on both. Local governments hold significant decentralized authorities in initiating local policy innovation, learning and adopting policies from other regions and implementing various policies. Without the cooperation and mobilization of local governments, the central government can hardly achieve serious SO_2 mitigation or any environmental cleanup.

However, much authority remains substantially centralized, especially setting up national goals for environmental protection. Various considerations for or against strong environmental protection are centrally weighed to form strong or weak political will by the top leadership of the Chinese Communist Party, as discussed in Chapter 2. It is then reflected in Five-Year Plans. When the top leadership determines to prioritize environmental protection among other governmental affairs, pollution mitigation goals started to enter as the key goals into National Five-Year Plans. These national goals are then decomposed into provincial goals for their implementation, as examined in Chapter 4. The goal allocation further penetrates into municipality and county levels, one level at a time. The types and stringency of goals closely follow the centralized political will for environmental protection. In addition, ministries and their internal departments in the central government are also directed by those goals to make policies and supervise provincial and other local governments for goal attainment. If a crucial environmental goal in the Five-Year Plan is missed, the Ministry of Ecology and Environment as the primary responsibility bearer will also be held accountable.

Credible mechanisms are established for central and local governments to make efforts for their goals. Most important, provincial leaders and ministers in the central government have their promotion opportunities controlled centrally through the Chinese Communist Party, while the fates of municipality-level leaders are determined at the provincial level. The clear linkages between their career development and goal-centered job performance are crucial incentives to motivate their genuine efforts but not just lip service.

Under goal-centered governance, it is the succession of goals but not individual policies that define environmental milestones. The top leadership of the party and the central government cares more about whether a certain goal has been achieved rather than a certain policy has been effective, efficient or fully implemented. In addition, the fact that China has not established a sound rule of law is also an important facilitating factor for enabling a goal-centered policy supply. Local leaders in charge, such as provincial governors, municipality mayors and county leaders as well as their corresponding party secretaries, will be less likely to lose their jobs or promotion opportunities for having policy failures, but the probability will increase significantly if a crucial goal does not get achieved. If one policy does not work, new ones will be quickly enacted to inch toward goal attainment.

2.2 Policy evolution by implementation selection

Policy supply under goal-centered governance has two key components: environmental goals to shape policy demand and low policy-making barriers and strong incentives to enable policy supply. More ambitious environmental goals will create stronger demand for pollution mitigation actions and thus a larger number of and more stringent policies. Goal-centered governance significantly reduces barriers for making policies. The much lower policy-making barriers result in intensive policy-making activities, competition among policies and much faster policy cycles. With a significant number of policies, each makes a small step toward an intended goal, although some are more important than others. The failure/success of any policy does not determine, but only to a limited extent affects, the final environmental outcome. Besides laws, a large number and wide variety of policies can be found in China on environmental protection that are enacted by various authorities, including the Central Committee of the Chinese Communist Party, the State Council, ministries and their composing departments (www.mee. gov.cn/zcwj/) and local governments.

Several causes contribute to the low policy-making barriers. The significantly decentralized policy making effectively reduces the barriers from the perspective of policy suppliers as they have a wide variety of sectoral and geographic jurisdictions and authorities. One consequence of this goal-centered policy supply is that it encourages policy innovation. Local governments have significant flexibility in deciding how to achieve top-down goals. Decentralized policy makers can weigh the significance, costs and benefits of various policies and their suitability to local contexts with dramatic regional disparities. Policies are constantly churned out from these decentralized policy makers at various levels to try their effectiveness in approaching goals. The effective mobilization of local governments not only facilitates policy enforcement, but it also creates incentives for even more active local environmental policy making if goal attainment so requires.

Furthermore, several key questions should be considered over the making of individual policies, while goal-centered governance has much lower requirements on policy designs to effectively decrease the policy-making barriers. First, how to ensure the quality of individual policies? Policies may be directly adopted from other countries and regions, revised to suit local contexts or innovated from scratch. China's colossal size and complexity indicate that many environmental policies can hardly be applied to fit all situations across the entire country. China's contexts are also sharply different from those in developed countries, where many environmental policies were first introduced and implemented. The decentralization of policy makers also indicates that the training and knowledge of those who write the policy texts may vary across local governments and ministries/departments. The much more greatly decentralized policy implementation and its unsatisfactory track record add further difficulties in understanding how policies could be designed better for more effective implementation. Accordingly, direct policy adoption is rarely effective, while policy localization and innovation are great challenges and require relevant knowledge and understanding. In addition, China's complexity also hinders timely-enough assessment of the crucial causes of any policy failure and success. Under goal-centered governance, the requirements on the quality of making individual policies are much lower because no policy or law occupies the central stage to solve a targeted environmental problem. The lower requirement for policy quality enables much swifter design and enactment processes. In other words, readers of China's environmental policies should not be primarily entangled in the enactment and effectiveness of individual policies, because they are of much less importance than goals. For example, essentially no SO_2 emission trading schedules have produced desirable outcomes that dominate SO_2 mitigation, but the failure had little impact on China's trajectory of controlling SO_2 emissions (Zhang et al., 2016).

Second, how to choose the most effective and efficient policy instrument among many alternatives? The choice of policy instruments is a crucial question for policy making, especially when a single or very few policies dominate the solutions to an environmental problem. Under goal-centered governance, this question is much less significant because policies are much less mutually exclusive. The considerably decentralized policy making also significantly reduces the possibility of any policy monopoly or oligopoly. The enactment of one policy instrument does not prevent the application of others. Accordingly, China does not need to choose a primary policy instrument for dealing with one environmental problem. For example, China's environmental protection tax law formally entered into force in January 2018, covering a wide variety of environmental pollutants, including SO_2 (National People's Congress, 2016). Many other crucial environmental policies are simultaneously in effect, such as the effluent emission standards that were examined earlier (MEP and AQSIQ, 2011).

Third, how are policies coordinated? While policies are individually made by different ministries and their internal departments, as well as various levels of governments, they can exert significant impacts on each other to create synergies and/or conflicts. Economic and energy policies are far beyond the jurisdiction of environmental protection. With local governments rather than their environmental protection bureaus in charge, coordination across these different types of policies became more feasible. In an optimized situation, policies should be well coordinated to maximize synergies and minimize conflicts. However, such coordination in China is inadequate in the context of decentralized policy making and especially policy implementation. Little evidence indicates that China rolls out the numerous policies for achieving the SO_2 mitigation goals in a systematic and coordinated way. Instead, the policy making is messy, with decentralized policy makers who have their individual authority in designing or shaping policies within their respective jurisdictions. Under goal-centered governance, however, such prior coordination of policy making is of lesser importance. After policies are made and put into implementation, they evolve rapidly. In China's context of weak rule of law and as examined earlier, individual policies have higher probabilities of unsatisfactory implementation. Similar to the natural selection process as proposed by Charles Darwin in understanding biological evolution (Darwin, 1859), policies in China also experience a dynamic evolution process and those fit ones are selected through implementation. Policies that have too many conflicts with others will be difficult to get effectively implemented. If one policy fails to achieve its intended consequences, new policies can be quickly introduced. Successful policies in one province can be rapidly adopted by other provinces or elevated to the national level.

Although much progress has been made in policy research in the past decade, such capacity was especially deficient in the early stages of SO_2 mitigation. China should still enhance its capability in policy making to improve the quality of individual policies, choose more wisely environmental policy instruments especially for those of relatively greater importance and scopes and better coordinate across policies. Nevertheless, the goal-centered policy supply substantially lowered the requirements for achieving desirable environmental protection outcomes such as serious mitigation of SO_2 emissions. The preceding crucial questions in policy making are of much less concern from their perspectives on influencing policy outcomes.

3 Policy scope for achieving SO_2 mitigation goals

China faces a wide scope in policy making for SO_2 mitigation. SO_2 emissions are affected by many economic, energy and environmental development factors and corresponding policies. Although the coal-fired power sector is increasingly important in coal consumption, still nearly two fifths of coal is consumed in other sectors (Figure 1.10). For achieving increasingly stringent SO_2 mitigation and environmental goals, the decentralized policy makers should evaluate the contributions of individual policies in policy supply.

3.1 Key factors for SO_2 emissions

SO_2 emissions can be decomposed with the following formula into various key factors:

$$SO_2 \text{ emissions} = GDP \times \frac{Energy}{GDP} \times \frac{Coal}{Energy} \times \frac{SO_2 \text{ emissions}}{Coal}$$

$$= GDP \times EI \times \frac{Coal}{Energy} \times \eta_s \times (1-\eta_{sr}) \times 2 \times (1-\eta_R)$$

Equation 5.1

"GDP" (gross domestic product) indicates the scale effect. Rapid economic growth in China leads to more SO_2 emissions. Energy consumption is a key foundation for any modern economy, and thus, energy intensity $\left(\dfrac{Energy}{GDP}\right)$ is another crucial effect. It measures how much energy is consumed for producing a given unit of GDP. Energy conservation and efficiency will reduce energy intensity and thus be beneficial for SO_2 mitigation. The economic structure also matters greatly. A greater proportion of service sectors in an economy could potentially reduce the overall energy intensity because in comparison to industrial sectors, they tend to consume much less energy for producing the same amount of economic outputs (Feng et al., 2009). China had a goal to reduce energy intensity by 20% in the 11th Five-Year Plan (National People's Congress, 2006). The Chinese central government also declared its intention in the 12th Five-Year Plan to "change the economic growth pattern," with a focus on energy conservation and environmental protection (National People's Congress, 2011). These two effects are related to economic development and energy conservation, on which economic and energy policies exert important influences.

Because coal consumption dominates the sources of SO_2 emissions, the share of coal in the energy mix is thus critical in deciding the sulfur intensity of energy.

$\dfrac{Coal}{Energy}$ is referred to as the energy transition effect. Its reduction is another measure for bringing down SO_2 emissions, which largely falls into the category of energy development and the scope of energy policy.

$\dfrac{SO_2 \; emissions}{Coal}$ refers to the mitigation effect, which is primarily decided by environmental policies. In combustion, a certain proportion of sulfur (η_{sr}) will be retained in ash and thus not emitted. This rate is mainly decided by the coal type and combustion technology, but not by policy intervention. Sulfur content in coal (η_s) is an important indicator of coal quality. The control of sulfur contents is often targeted in early environmental regulations for reducing SO_2 emissions. SO_2 scrubbers and other SO_2 removal measures can avoid a certain share of SO_2 (η_R) from being emitted after generation.

China's economy has been growing at an astonishing pace in the past four decades. Real GDP in 2018 was 31.7 times of that in 1980 with a growth rate of 9.5% annually, while real GDP per capita rose to be 22.4 times or 8.5% annually (Figure 5.1). As measured in nominal GDP of current U.S. dollars, China overtook Japan to become the second-largest economy in the world in 2010 and further rose to be equivalent to 65.0% of the United States in 2018 (Figure 5.1). China's much larger population indicates that the country's GDP per capita still trails the global average and is a small fraction of that in Japan and the United States. Although the GDP growth rate has been significantly slower in the 2010s than in the 2000s,

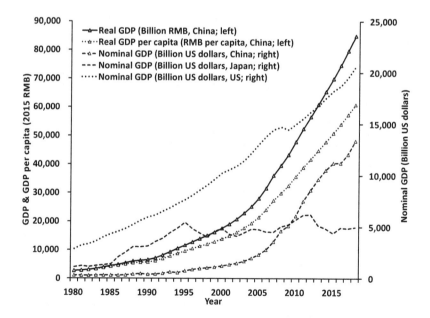

Figure 5.1 Economic growth in China, Japan and the United States

Source: IMF (2019).

the convergence of average living standards in China toward that of developed countries is expected to further intensify economic activities within its geographical territory and thus to add great environmental pressures.

Energy consumption is not only one key foundation for economic development, but it also brings unwanted consequences of environmental pollution. The combustion of fossil fuels, especially coal, is the primary source of air pollutant emissions that cause ambient particulate matter (PM) pollution. Although China has been improving its energy efficiency for producing one unit of GDP especially in the past decade, its primary energy consumption climbed up quickly. When consuming one ton of oil equivalent of primary energy, China in 2018 produced US$4,084 of nominal GDP, while the rates for Japan and the United States were US$10,948 and US$8,945, respectively (IMF, 2019; BP, 2019). Due to the significantly lower energy efficiency, China overtook the United States to become the largest energy consumer in the world in 2009, but its economy then was two thirds smaller. A major shift took place in around 2003, and since then, China's energy consumption has been growing at a much faster pace than before (Figure 5.2). Not only China's economic growth accelerated after 2003, but also the energy efficiency reversed its earlier improvement trend to decrease between 2002 and 2005 (Figure 5.2). In 2018, China consumed 224% more primary energy than in 2000 to become 42% higher than the United States' level (Figure 5.2).

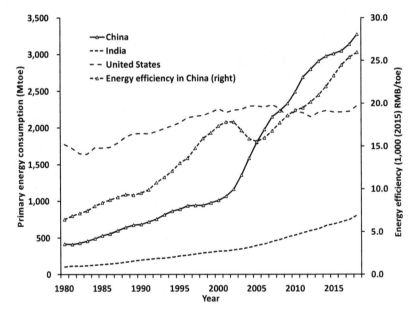

Figure 5.2 Primary energy consumption and energy efficiency

Source: BP (2019).

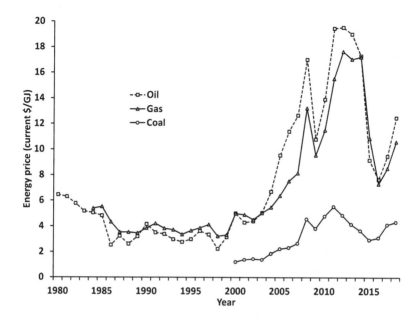

Figure 5.3 Prices of coal (Qinhuangdao spot price), oil and natural gas
Source: Japan LNG CIF; BP (2019).

China's low level of GDP per capita might partly explain why its energy mix heavily focuses on coal. As shown in Figure 5.3, coal is the cheapest and most affordable among the three major fossil fuels. When China's economy grew very fast, especially in the 2000s, to rapidly push up energy consumption, coal became the primary choice for meeting the additional energy demand (Figure 5.4), and thus, its share in the energy mix even reversed its earlier declining trend to become higher in the early 2000s (Figure 5.5).

In the past decade, energy transition has also been playing an increasingly visible role that led to the mitigation of SO_2 emissions. China's energy mix is heavily tilted toward coal, the most pollution-intensive fuel. With more coal burning squeezed into China's territory, the pressure on the environment is mounting. Throughout the 1980s and 1990s, the share of coal was continuously above 70% (Figure 5.5). The slow declining trend in the 1990s was reversed in early 2000s to witness the share climbing up again from 69.5% in 2001 to 73.7% in 2007, further intensifying environmental pollution in China. The following decade witnessed an unprecedented decrease and coal's share had dropped to 58.2% in 2018. Nevertheless, China still accounted for 50.5% of global coal consumption in 2018 (BP, 2019). Although oil and natural gas have increasing shares in China's primary energy consumption, the overall share of fossil fuels experienced an accelerated decline from 94.1% in 2007 to 85.3% in 2018. Nonfossil fuels are much more

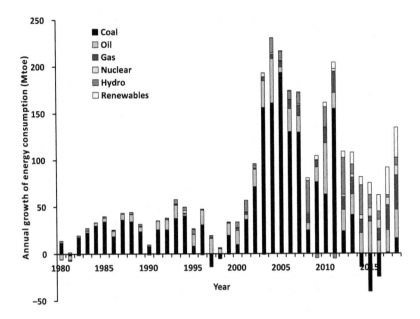

Figure 5.4 The annual growth of primary energy consumption in China by fuels
Source: BP (2019).

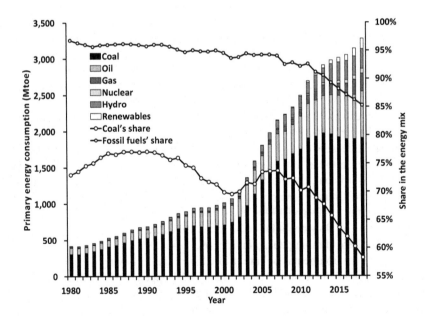

Figure 5.5 China's primary energy consumption by fuel and the shares of coal and fossil fuels
Source: BP (2019).

important in the energy mix, from 5.9% in 2007 to 14.7% in 2018 (Figure 5.5). Nuclear, hydropower and nonhydro renewables witnessed their shares increased from 0.7%, 5.1% and 0.2% in 2007 to 2.0%, 8.3% and 4.4% in 2018, respectively. Nonhydro renewables were the fastest-growing energy type.

As one indicator of energy modernization, China's primary energy consumption is rapidly electrifying to reshape the major sources and sectors of SO_2 emissions. In 1990, only 20.8% of primary energy consumption went through the intermediate stage of electricity before final consumption, which was only slightly higher than Africa's 17.8% (Figure 5.6). With rapid energy modernization, this ratio increased to 42.5% in 2015, then similar to the United States' 40.3% (Figure 5.4). Rapid electrification also happened in other rapidly industrializing countries such as India, but the progress in Africa has been much slower (Figure 5.4). With China's continuous efforts for electrifying energy consumption – such as the push for electric vehicles (IEA, 2019) – this electrification rate is expected to further escalate, which will distinguish the importance of the power sector in China's energy consumption and environmental protection.

Energy transition for electricity generation is even more visible. Coal's share has been reduced significantly from the 81.0% peak in 2007 to 66.5% in 2018. Other fossil fuels, including oil and natural gas, accounted for only an insignificant share at 3.3% in 2018 (Figure 5.7). In contrast, the share of nonhydro renewables, mostly wind and solar energy, has achieved the largest growth from 0.5% in 2007 to 8.9% in 2018 (Figure 5.7). Coal's share in electricity generation is significantly

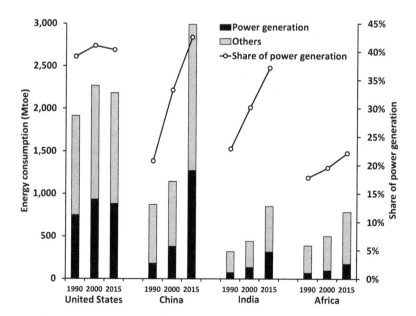

Figure 5.6 Primary energy consumption and its electrification rate

Source: IEA (2017).

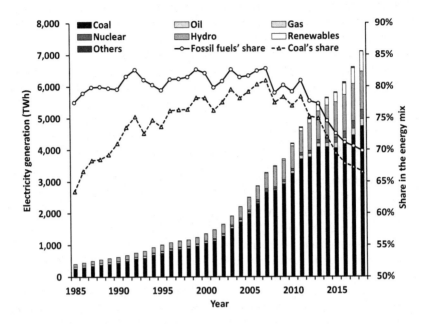

Figure 5.7 Electricity generation by fuels in China
Source: BP (2019).

higher than that in primary energy consumption, being 66.5% and 58.2% in 2018, respectively (Figure 5.8). From 2007 to 2018, their drops were 14.4% and 15.4% in percentage points, respectively. Nonfossil fuels – such as nuclear, hydro and nonhydro renewables – are generally for electricity generation, while oil and natural gas in China are primarily consumed not in the power sector. Especially in the past decade, the advancement of renewables significantly accelerated to account for increasingly sizable shares of electricity generation growth (Figure 5.8).

The power sector has been increasing its importance in coal consumption. In 1980, only 20.2% of China's coal consumption was in the power sector, while this ratio climbed steadily to 52.2% in 2002 before a decade-long stabilization (Figure 1.10). During 2015~2017, the increasing trend restarted to reach 57.3% in 2017 from 50.3% in 2014 (Figure 1.10). This ratio is expected to further increase, in reference to the situation in the United States, whose power sector accounted for 18.6% of coal consumption in 1950 and 92.8% in 2017 (Figure 1.10). The trend indicates that the energy mix in nonpower sectors shifts away from direct coal consumption faster than that in the power sector, although the former may consume more electricity that comes from coal-fired power plants.

In China's trajectory of SO_2 mitigation, these economic, energy and environmental factors made different contributions in different Five-Year Plans (Figure 5.9). SO_2 emissions went down by 15.8% in the 9th Five-Year Plan

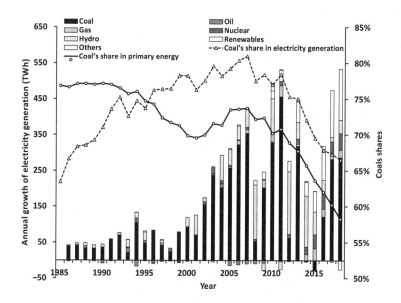

Figure 5.8 The annual growth of electricity generation in China by fuels and coal's share
Source: BP (2019).

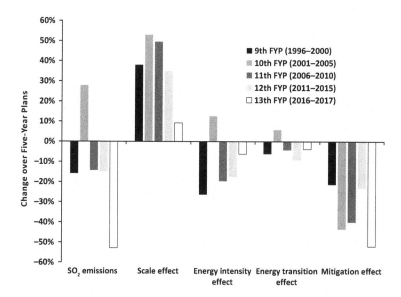

Figure 5.9 The decomposition of China's SO$_2$ emissions
Source: National Statistics Bureau and Ministry of Ecology and Environment (2019); IMF (2019); BP (2019).
Note: Method comes from Ang (2005).

(1996–2000). Under the influence of the Asian financial crisis of 1997, the scale effect would still lead to an increase of 38.0%, while the effects of energy intensity, energy transition and mitigation reduced SO_2 emissions by 26.4%, 6.1% and 21.3%, respectively, over the five years (Figure 5.9). They indicated the varying impacts of economic, energy and environmental policies and development. Specifically, as represented in the mitigation effect, environmental policies made an important but certainly not decisive contribution. The 10th Five-Year Plan had a very different picture. With accelerated economic growth, the scale effect would boost emissions by 53.0%, while the energy intensity and energy transition effects also pushed the emissions upward by 12.6% and 5.7%, respectively. Although the mitigation effect of 43.5% reduction was much greater than that in the 9th Five-Year Plan, the outcome was that SO_2 emissions increased by 27.8%. In other words, the deterioration was not due to less effective environmental policies, but faster economic expansion reversed trends of energy intensity and transition.

Reversing the deterioration trend, the 11th Five-Year Plan managed to reduce SO_2 emissions by 14.3%. The following 12th Five-Year Plan registered a similar reduction of 14.9%. The four effects also had comparable contributions in these two Five-Year Plans: the scale effect 49.6% versus 35.0%, the energy intensity effect −19.6% versus −17.6%, the energy transition effect −4.2% versus −9.1% and the mitigation effect −40.0% versus −23.2% (Figure 5.9). In the first two years (2016–2017) of the 13th Five-Year Plan with available data, SO_2 emissions dropped by whopping 52.9%, and the mitigation effect contributed decisively a reduction of 52.0% (Figure 5.9).

3.2 Technological factors for effluent SO₂ emissions in the power sector

Electrification of energy consumption and the power sector's increasing share in coal consumption distinguish the importance of coal-fired power plants in controlling China's SO_2 emissions. Given the understanding of coal combustion and SO_2 emissions in electricity generation, the SO_2 emission intensity $\left(\dfrac{SO_2 \ emissions}{Coal} \right)$ can be converted to effluent SO_2 concentration, mg/Nm³ (Ministry of Environmental Protection and General Administration of Quality Supervision Inspection and Quarantine, 2011). The SO_2 concentration is measured under normal conditions (thus "N"), with excess air coefficient being 1.4. The value 1.4 here indicates that 40% more air or, specifically, oxygen will be blown into boilers than what is required for complete combustion. Excess air is needed for more complete combustion within a short residence time of fuels in boilers, but excess air will also take heat away and lower the thermal efficiency. Accordingly, an optimum value exists, not necessarily being 1.4 for every power plant. The fixed value is for policy purposes and intends to prevent cheating because a convenient option of lowering effluent SO_2 concentration is to dilute the flue gas with more air. Approximately, excess air coefficient α can be calculated as $\alpha \approx \dfrac{21\%}{21\% - x\%}$, $x\%$ referring to the percentage of O_2 in flue gas, when $\alpha = 1.4$ and $x\% \approx 6\%$.

As revealed in Equation 5.1, there are three key technological factors to decide SO_2 emission intensity and effluent SO_2 concentration. The first factor is the fractions of sulfur retained in ash (η_{sr}). When coal is burned in boilers, sulfur is converted into several forms, being gaseous (SO_2, SO_3, gaseous sulfates) and solid (in bottom ash and as particulate sulfate; EPA, 1998). SO_2 greatly dominates the gaseous forms (EPA, 1998). Higher combustion temperature leads to lower fractions of sulfur retained in ash, which disadvantages pulverized coal (PC) combustion against fluidized bed combustion (FBC; Sheng et al., 2000). Another factor is the calcium/sulfur (Ca/S) molar ratio in coal: a higher Ca/S ratio facilitates sulfur retention (Cheng et al., 2004; EPA, 1998). (The Ca/S ratio here is different from the Ca/S ratio for SO_2 scrubbers as discussed later.) For PC combustion, the presence of calcium is much less important than FBC due to the thermal instability of calcium sulfate ($CaSO_4$), the main product responsible for sulfur retention (Sheng et al., 2000). The fractions of sulfur retained in the application are summarized in Table 5.1. Compared with China's assumption of 20% (State Council, 2007a), fractions of sulfur retained are widely believed to be significantly lower, except for the situations of burning lignite and applying FBC technologies. Even another Chinese official document believed the rate to be 10% to 15% and recommended 10% for the purpose of designing SO_2 scrubbers (NDRC, 2004). China's SO_2 emissions could have been underestimated partly because of the choice of this parameter (Figure 1.8).

Second, lower sulfur contents in coal are crucial for reducing SO_2 generation intensity. An official data set is employed to analyze the distribution of sulfur contents for SO_2 scrubbers. In the 11th Five-Year Plan on Acid Rain and SO_2 Pollution Control, China published data for 248 coal power plants with a total capacity of 164 GW that covered all SO_2 retrofit projects to be completed between 2006 and 2010 (SEPA and NDRC, 2008). Sulfur contents estimated from this data

Table 5.1 Applied fractions of sulfur retained in ash

Coal type	Fractions of sulfur retained in ash	Source	
Bituminous, PC*	5%	U.S. Environmental Protection	(EPA, 1998)
Sub-bituminous, PC*	12.5%	Agency's (EPA's) choice	
Lignite, PC*	25%		
Coal in general	≤10%	U.S.'s study	(Singer, 1981)
Nonlignite coal	5%	Assumption from the U.S. EPA	(Smith et al., 2001)
Lignite	30%	Assumption in the research	
Coal	5%~30%	The study's assumption	(Ohara et al., 2007)
Coal	5%~10%	China's study	(Zhao et al., 2008)
Coal, PC*	10%~15%	China's official recommendation for scrubber design	(NDRC, 2004)
Coal	20%	Assumption in compiling China's statistical data	(State Council, 2007a)

Note: PC* refers to pulverized coal power plants.

set are expected to represent their distribution for all coal power plants with SO$_2$ scrubbers. Each plant disclosed information on scale (MW), year, annual SO$_2$ removal capability (tons per year), location and name. Sulfur contents can then be estimated under the following assumptions: thermal efficiencies were 370 g of coal equivalent per kilowatt-hour, or 1,930 kWh per ton coal (the average efficiency in 2005 [China Electricity Council, 2006–2015]); capacity factors were 5,500 hours per year (SEPA, 2006a); 80% of the sulfur was converted to SO$_2$ in combustion and 20% was retained in ash, as recognized in China's official statistics (State Council, 2007a); and overall SO$_2$ removal rates were 85% (SEPA, 2007). The calculation formula is

$$\text{Sulfur content} = \frac{\text{SO}_2 \text{ removal capability}}{\dfrac{\text{Coal power capacity} \times 5500}{1930} \times 2 \times 80\% \times 85\%}$$

"2" in the denominator refers to the fact that when sulfur is converted to SO$_2$, the mass doubles since the molecular weight of SO$_2$ is twice that of sulfur. A caveat is that these numbers are used here to reversely calculate sulfur contents because they represent China's original assumption in compiling the data. One legitimate concern is the accuracy of the assumed 80% conversion. Actually, in the combustion of anthracite, bituminous and sub-bituminous coal, 90% or more of the sulfur is converted to SO$_2$, as discussed earlier. In addition, as discussed in Chapter 6, China's actual SO$_2$ removal rates should be significantly lower than 85% especially before 2007. Actual thermal efficiencies and capacity factors also vary across years.

Sulfur contents are closely related to the costs of SO$_2$ mitigation. Generally speaking, higher sulfur contents correspond to lower costs for every ton of SO$_2$ removed but higher costs for every kilowatt-hour of electricity generated. The distribution of sulfur contents is shown in Figure 5.10 with the national average being about 1.0%. Of the coal-fired power plants, 68% burned coal with less than 1% sulfur and another 26% between 1% and 2%. The remaining 6% of the total capacity was associated with higher than 2%-sulfur coal. China not only installed SO$_2$ scrubbers not only in coal power plants burning high-sulfur coal but also in those burning low-sulfur coal. China's distribution of sulfur contents had a single peak at around 0.75% (Figure 5.10), which reflected the fact that most of China's coal is mined in one region. For example, two thirds of China's coal production in 2007 came from the seven nearby provinces of Shanxi, Inner Mongolia, Shaanxi, Shandong, Anhui, Hebei and Henan (National Bureau of Statistics, 1997–2008).

The preceding two factors decide how much SO$_2$ is generated when burning a unit quantity of coal. SO$_2$ removal rates are the third factor to reduce the SO$_2$ emission intensity. Before construction begins, a report of environmental impact assessment (EIA) had to be submitted to a governmental authority on environmental protection (NPC, 2002). If the plant was believed to bring unacceptable environmental damage – for example, seriously worsen ambient air quality – the

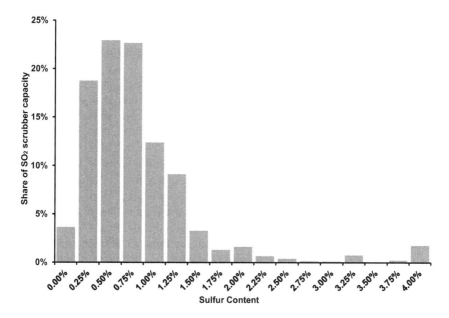

Figure 5.10 Distribution of sulfur contents in coal power plants in China (with retrofitted SO₂ scrubbers)

Source: SEPA and NDRC (2008).

EIA report would be rejected. Another policy – "three simultaneities" – required pollution control facilities to be designed, constructed and completed at the same time as the main project (State Council, 1998). For example, if SO_2 scrubbers were considered necessary in the EIA report, this policy would demand their installation.

3.3 Technical measures for SO_2 removal in the power sector

In order to remove SO_2 in electricity generation, coal-fired power plants in China are required to meet effluent emission standards that are made more stringent every six or seven years to reflect growing environmental concerns. In the standards enacted in 1996, new coal-fired power plants that passed EIA after January 1997 should achieve 2,100 mg/Nm³ (if burning coal with ≤1% sulfur) or 1,200 mg/Nm³ (if burning coal with >1% sulfur; SEPA and AQSIQ, 1996). For coal-fired power plants burning bituminous coal with 0.5% sulfur, SO_2 concentration in the non-desulfurized flue gas will generally exceed 1,000 mg/Nm³. Essentially the 1996 effluent emission standards meant that coal-fired power plants burning coal with >1% sulfur should have SO_2 scrubbers while those with ≤1% sulfur did not need to. In the standards enacted in 2003, for the great majority of coal power plants, their effluent SO_2 concentration should be kept below 400 mg/Nm³ on 1 January 2010 (SEPA and General Administration of Quality Supervision Inspection and Quarantine, 2003). In addition to China's shutting down old, small power-generating units, the effluent emission standard itself would ensure that a dominant share of China's coal power capacity in 2010 would have SO_2 scrubbers installed and operate normally.

The standards were updated in 2011 for being effective on 1 January 2012 (Ministry of Environmental Protection and General Administration of Quality Supervision Inspection and Quarantine, 2011). New plants should then reduce their effluent SO_2 emissions below 100 mg/Nm^3 while the standard for existing plants was 200 mg/Nm^3. In southwestern provinces, including Guangxi, Chongqing, Sichuan and Guizhou, where local coal contains much higher sulfur contents, the standards could be relaxed to 200 mg/Nm^3 and 400 mg/Nm^3, respectively. Natural gas–fired power plants tend to be much cleaner, with the standard being 35 mg/Nm^3.

In 2014, a new policy, "Upgrading and Retrofitting Action Plan for Energy Conservation and Pollution Mitigation in the Coal-Fired Power Sector," was enacted jointly by National Development and Reform Commission, Ministry of Environmental Protection and National Energy Administration (National Development and Reform Commission et al., 2014). It *required* newly constructed coal-fired power plants in eastern provinces to achieve the standard for natural gas–fired power plants, that is, 35 mg/Nm^3 for SO_2 emissions. Central provinces should *approach* this standard, while western provinces were *encouraged* to reach the level. This much more stringent standard is referred to in China as the *ultra-low* emissions. In 2015, another policy mandates the ultra-low standard to be achieved in most new and existing coal-fired power plants with only occasional exceptions (Ministry of Environmental Protection et al., 2015).

China's Law of Standardization and its implementation regulations provide legal teeth (NPC, 1988; State Council, 1990). Effluent emission standards are clearly stated as "mandatory standards" (State Council, 1990), while products not meeting "mandatory standards" are forbidden to produce, sell and import (NPC, 1988). In this sense, coal power plants should stop generating electricity if the effluent SO_2 emissions exceeded corresponding standards. The electric grid should not accept the electricity if it were not legally generated.

In order to achieve SO_2 removal rates as required by the stringent ultra-low effluent emission standard, coal-fired power plants should generally achieve very high SO_2 removal rates, being 98.5% if Huolinhe lignite or Datong bituminous coals are burned or 96.9% for Shenfu bituminous coal (Table 5.2). The SO_2 emission intensity of electricity generation should also be substantially reduced to about 0.10 to 0.11 g/kWh. The sulfur contents in these three types of coal, from 0.50% to 0.99%, fall within the normal range. For high-sulfur coal, especially in southwestern provinces, the required SO_2 removal rates are much higher, generally beyond 99%. The deep reduction can only be achieved through SO_2 scrubbers if coal remains as the fuel.

Before China started the large-scale deployment of SO_2 scrubbers in the early 2000s, the world in total had installed about 200 GW (Taylor et al., 2005). The United States accumulated around 100 GW in a 25-year period between 1975 and 2000 (Taylor et al., 2005). Germany and Japan together took 30% of the world's

Table 5.2 Effluent SO₂ emissions and necessary SO₂ removal rates

		Huolinhe lignite	*Datong bituminous*	*Shenfu bituminous*
LHV (MJ/kg)		13.9	21.0	21.4
Contents in coal (%)	Sulfur	0.61%	0.99%	0.50%
	Carbon	34.1%	55.7%	57.0%
	Hydrogen	2.7%	3.4%	3.4%
	Oxygen	10.5%	8.3%	8.0%
	Nitrogen	0.7%	0.9%	1.1%
Effluent SO₂ emissions (without removal)	Concentration (mg/Nm³)	2,315	2,291	1,133
	Emissions (g/kWh)	6.79	7.27	3.60
For achieving the 35 mg/Nm³ standard	Required SO₂ removal rate (%)	98.5%	98.5%	96.9%
	Emissions (g/kWh)	0.10	0.11	0.11

Note: Assuming the sulfur retention rate in ash, 90%; thermal efficiency of electricity generation (42%, or 293 g of coal equivalent/kWh). Coal quality data are from Shi and Yu (2005).

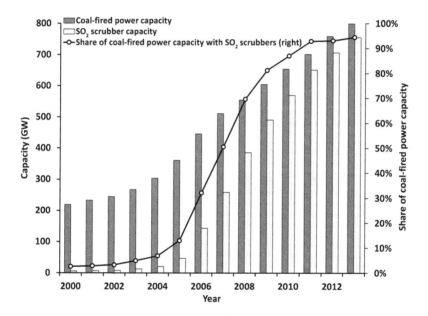

Figure 5.11 Coal-fired power and SO₂ scrubber capacities in China

Source: Ministry of Environmental Protection (2014); EIA (2019); China Electricity Council (2010, 2006–2015).

market, and the remaining 20% were in other countries (Taylor et al., 2005). The scrubber capacity numbers presented in Figure 5.11 were calculated from a publicly available plant-level data set (Ministry of Environmental Protection, 2014). The dataset includes information on the name and location of coal power plants, the serial number and power capacity of generators, the dates that generators and

SO$_2$ scrubbers came online, SO$_2$ scrubber technology type and the name of the SO$_2$ scrubber company in charge. The SEPA (Ministry of Environmental Protection after March 2008) established a standard procedure for registering SO$_2$ scrubbers (SEPA, 2005, 2006b); for example, an SO$_2$ scrubber had to operate continuously for 168 hours to test its performance before registration.

China's share was negligible with only 5.6 GW of SO$_2$ scrubbers at the end of 2000, or 2.5% of its coal-fired power capacity (Figure 5.11). In the 10th Five-Year Plan (2001–2005), SO$_2$ scrubber capacity rose to 46.7 GW in 2005. The progress was noticeable and the proportion of coal-fired power capacity with SO$_2$ scrubbers increased to 12.5%. However, because total coal-fired power capacity escalated from 218.9 GW in 2000 to 360.6 GW in 2005, essentially China had more coal-fired power plants without SO$_2$ scrubbers to witness a steady increase of the power sector's SO$_2$ emissions. In the 11th Five-Year Plan (2006–2010), coal-fired power capacity grew at a much faster pace to reach 654.3 GW in 2010, while SO$_2$ scrubber capacity was lifted even faster to 569.3 GW in 2010. Then only 85.0 GW of coal-fired power plants, or 13.0%, did not have SO$_2$ scrubbers. This rate had already been much lower than that in 2000. In the following years, the ratio further inched higher to 94.4% in 2013. Essentially, in China, nearly all coal-fired power plants should have SO$_2$ scrubbers to continue operation.

The 11th Five-Year Plan witnessed a large-scale campaign to retrofit existing coal-fired power plants (Figure 5.12), besides shutting down many inefficient

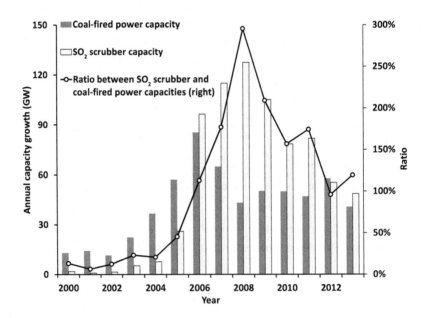

Figure 5.12 The annual growth of coal-fired power and SO$_2$ scrubber capacities in China

Source: Ministry of Environmental Protection (2014); EIA (2019); China Electricity Council (2006–2015, 2010).

small units (Xu et al., 2013). The ratios between the annually increased capacities of SO_2 scrubbers and coal-fired power plants were consistently higher than 100% in every year of the 11th Five-Year Plan. At the retrofitting peak in 2008, SO_2 scrubber capacity grew by 127.4 GW while coal-fired power capacity increased only by 43.1 GW. After the 12th Five-Year Plan, a great majority of SO_2 scrubbers were either built together with coal-fired power plants or further retrofitted for meeting more stringent effluent emission standards.

Most SO_2 scrubbers fall into three scale categories: 300 MW, 600 MW and 1000 MW (Figure 5.13). They correspond to several predominant, standardized unit scales that China's coal-fired power units have. Among the 754.9 GW of coal-fired power units with SO_2 scrubbers in 2013, 62.1 GW, 215.1 GW and 263.9 GW were within the 1,000~1,050-MW, 600~650-MW and 300~350-MW ranges, respectively. Two smaller scales have seen their importance fading after China focused more on larger and more efficient units. Respectively, 42.3 GW and 32.7 GW fell within the 200~220-MW and 135~150-MW ranges. In total, these five standardized unit sizes accounted for 618.6 GW or 81.9% of all SO_2 scrubbers. These size and technology standardization provided one crucial advantage in designing and rapidly deploying SO_2 scrubbers.

The geographic distribution of SO_2 scrubbers reflects that of coal-fired power plants. Provinces in East China, North China and South China had 249.2 GW, 213.0 GW and 104.1 GW (or 33.0%, 28.2% and 13.8%) of SO_2 scrubbers,

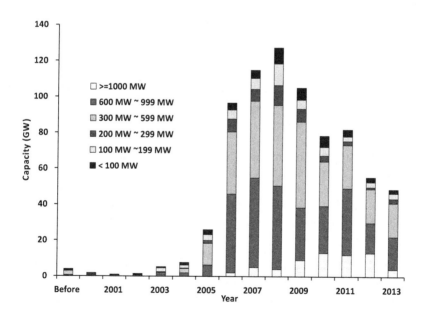

Figure 5.13 Annually increased SO_2 scrubber capacity and unit sizes

Source: Ministry of Environmental Protection (2014).

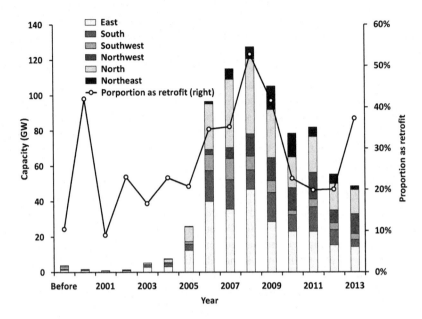

Figure 5.14 The annual growth of SO$_2$ scrubber capacity by regions (as categorized by the six Regional Supervision Bureaus of the Ministry of Ecology and Environment; SO$_2$ scrubbers are called "retrofits" when the online dates of SO$_2$ scrubbers and coal power units are over one year)

Source: Ministry of Environmental Protection (2014).

respectively, in 2013 (Figure 5.14). Northeast, Southwest and Northwest had 188.5 GW in total, or 25.0%. Their vast geographic territories indicate that these coal-fired power plants are scattered at much greater distances from each other to potentially enhance difficulties for environmental compliance monitoring and enforcement.

Different SO$_2$ scrubber technologies correspond to a wide range of possible SO$_2$ removal rates. China had 1589 units of SO$_2$ scrubbers at or above 200 MW in 2013. The limestone-gypsum wet type is the most applied technology especially for large coal-fired power units, accounting for 93.6% of units >=1000 MW, 96.1% of those between 600 MW and 999 MW, 87.6% of those between 300 MW and 599 MW and 81.0% of those between 200 MW and 299 MW (Figure 5.15). The share dropped significantly for units smaller than 200 MW, being only 30.0% (Figure 5.15). Due to the same consideration of economy of scale, seawater type also heavily tilted toward large units (Figure 5.15). Only 94.5 GW of SO$_2$ scrubbers in 2013 were individually smaller than 200 MW (12.5% of all SO$_2$ scrubbers), but they had 2,878 units (64.4% of all; Figure 5.15).

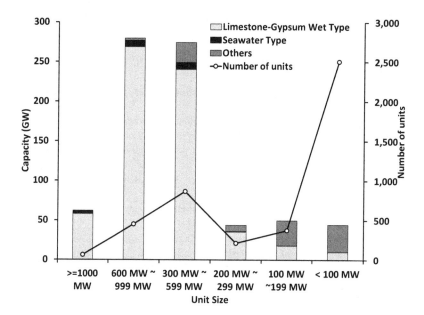

Figure 5.15 SO$_2$ scrubbing technologies by unit sizes
Source: Ministry of Environmental Protection (2014).

Reference

Ang, B. W. 2005. The LMDI approach to decomposition analysis: A practical guide. *Energy Policy*, 33, 867–871.

Barron, W. F. & Ng, G. T. L. 1996. An assessment methodology for environmental policy instruments: An illustrative application to solid wastes in Hong Kong. *Journal of Environmental Management*, 48, 283–298.

Benkovic, S. R. & Kruger, J. 2001. US sulfur dioxide emissions trading program: Results and further applications. *Water Air and Soil Pollution*, 130, 241–246.

Blackman, A. 2010. Alternative pollution control policies in developing countries. *Review of Environmental Economics and Policy*, 1–20.

BP. 2019. *Statistical review of world energy* [Online]. Available: www.bp.com/en/global/corporate/energy-economics/statisticalreview-of-world-energy.html.

Cheng, J., Zhou, J. H., Liu, J. Z., Cao, X. Y., Zhou, Z. J., Huang, Z. Y., Zhao, X. & Cen, K. 2004. Physicochemical properties of Chinese pulverized coal ash in relation to sulfur retention. *Powder Technology*, 146, 169–175.

China Electricity Council. 2006–2015. *Annual report of national power generation*. Beijing, China: China Electricity Council.

China Electricity Council. 2010. *The basic information of electricity generation units and electric grids*. Beijing, China: China Electricity Council.

Darwin, C. 1859. *On the origin of species by means of natural selection, or the preservation of favoured races in the struggle for life*. London: J. Murray.

Dong, Z., Ge, C., Gao, S. & Wang, J. 2011. Assessment on the Chinese pollution charges system and proposals to the future reform. *In:* Shen, M., Ge, C., Dong, Z. & Zhang, B. (eds.) *Progress on environmental economics*. Beijing, China: China Environmental Science Press.

EIA. 2019. *International energy statistics*. Washington, DC: U.S. Energy Information Administration.

EPA. 1998. *AP 42, chapter 1: External combustion sources*, fifth ed. Washington, DC: EPA.

Feng, T. W., Sun, L. Y. & Zhang, Y. 2009. The relationship between energy consumption structure, economic structure and energy intensity in China. *Energy Policy*, 37, 5475–5483.

Ge, C., Ren, Y., Gao, S., Sun, G. & Long, F. 2011. Wastewater discharge tax design: From pollutant discharge levy to environmental tax. *In:* Shen, M., Ge, C., Dong, Z. & Zhang, B. (eds.) *Progress on environmental economics*. Beijing, China: China Environmental Science Press.

Goulder, L. H. & Parry, I. W. H. 2008. Instrument choice in environmental policy. *Review of Environmental Economics and Policy*, 2, 152–174.

IEA. 2017. *World energy outlook*. Paris, France: IEA.

IEA. 2019. *Global EV outlook 2019*. Paris, France: IEA.

IMF. 2019. *World economic outlook database October 2019*. Washington, DC: IMF.

MEP & AQSIQ. 2011. *Emission standard of air pollutants for thermal power plants*. Beijing, China: MEP & AQSIQ.

Ministry of Environmental Protection. 2014. *The list of China's SO_2 scrubbers in coal-fired power plants*. Beijing, China: Ministry of Environmental Protection.

Ministry of Environmental Protection & General Administration of Quality Supervision Inspection and Quarantine. 2011. *Emission standard of air pollutants for thermal power plants*. GB 13223-2011. Beijing, China: Ministry of Environmental Protection.

Ministry of Environmental Protection, National Development and Reform Commission & National Energy Administration. 2015. *Comprehensive implementation of the upgrading and retrofitting action plan for energy conservation and pollution mitigation in the coal-fired power sector*. Beijing, China: Ministry of Environmental Protection.

National Bureau of Statistics. 1997–2008. *China energy statistical yearbook*. Beijing, China: China Statistics Press.

National Development and Reform Commission, Ministry of Environmental Protection & National Energy Administration. 2014. *Upgrading and retrofitting action plan for energy conservation and pollution mitigation in the coal-fired power sector (2014–2020)*. Beijing, China: Ministry of Environmental Protection.

National People's Congress. 2001. *The outline of national 10th five-year plan on economic and social developments*. Beijing, China: The 4th Conference of the 9th National People's Congress.

National People's Congress. 2006. *The outline of the national 11th five-year plan on economic and social development*. Beijing, China: The 4th Conference of the 10th National People's Congress.

National People's Congress. 2011. *The outline of the national 12th five-year plan on economic and social development*. Beijing, China: The 4th Conference of the 10th National People's Congress.

National People's Congress. 2016. *Environmental protection tax law of the people's republic of China*. Beijing, China: The 4th Conference of the 10th National People's Congress.

National Statistics Bureau & Ministry of Ecology and Environment. 2019. *China statistical yearbook on environment 2018*. Beijing, China: China Statistics Press.

NDRC. 2001. *Key special 10th five-year plan on energy development*. Beijing, China: NDRC.

NDRC. 2004. *Technical code for designing flue gas desulfurization plants of fossil fuel power plants*. DL/T5196-2004. Beijing, China: NDRC.

NDRC. 2007. *11th five-year plan on energy development*. Beijing, China: NDRC.

NPC. 1988. *The standardization law of the people's republic of China*. Beijing, China: 7th National People's Congress of the People's Republic of China.

NPC. 2002. *Environmental impact assessment law of the people's republic of China*. Beijing, China: 9th National People's Congress of the People's Republic of China.

Ohara, T., Akimoto, H., Kurokawa, J., Horii, N., Yamaji, K., Yan, X. & Hayasaka, T. 2007. An Asian emission inventory of anthropogenic emission sources for the period 1980–2020. *Atmospheric Chemistry and Physics*, 7, 4419–4444.

Portney, P. R. & Stavins, R. N. 2000. *Public policies for environmental protection*. Washington, DC: Resources for the Future.

SEPA. 2001. *National 10th five-year plan on environmental protection*. Beijing, China: State Environmental Protection Administration.

SEPA. 2005. *Flue gas limestone/lime – gypsum desulfurization project technical specification of thermal power plant*. Beijing, China: State Environmental Protection Administration.

SEPA. 2006a. *Guidelines on calculating SO_2 emission quotas*. Beijing, China: State Environmental Protection Administration.

SEPA. 2006b. *Technical guidelines for environmental protection in power plant capital construction project for check and accept of completed project*. Beijing, China: State Environmental Protection Administration.

SEPA. 2007. *Detailed methods to verify major pollutants emission reduction in the 11th five-year period (on trial)*. Beijing, China: State Environmental Protection Administration.

SEPA & AQSIQ. 1996. *Emission standard of air pollutants for thermal power plants*. GB 13223-1996. Beijing, China: SEPA, AQSIQ.

SEPA & NDRC. 2008. *National 11th five-year plan on acid rain and SO2 pollution control*. Beijing, China: State Environmental Protection Administration, NDRC.

Sheng, C. D., Xu, M. H., Zhang, J. & Xu, Y. Q. 2000. Comparison of sulphur retention by coal ash in different types of combustors. *Fuel Processing Technology*, 64, 1–11.

Shi, D. & Yu, C. 2005. *Optimization design, technical control and coal quality evaluation of contemporary coal pre-processing*. Beijing, China: China Contemporary Audio and Video Press.

Singer, J. G. 1981. *Combustion, fossil power systems: A reference book on fuel burning and steam generation*. Windsor, CT: Combustion Engineering.

Smith, S. J., Pitcher, H. & Wigley, T. M. L. 2001. Global and regional anthropogenic sulfur dioxide emissions. *Global and Planetary Change*, 29, 99–119.

State Council. 1990. *Regulations for the implementation of the law of standardization*. Beijing, China: State Council.

State Council. 1998. *Administrative rule on environmental protection of construction projects*. Decree No. 253. Beijing, China: State Council.

State Council. 2007a. *Notice on distributing implementation plans and methods of statistics, monitoring and assessment on energy conservation and pollutant emission reduction*. Beijing, China: State Council.

State Council. 2007b. *Notice on distributing the 11th five-year plan on environmental protection*. Beijing, China: State Council.

State Environmental Protection Administration (SEPA) & General Administration of Quality Supervision Inspection and Quarantine. 2003. *Emission standard of air pollutants for thermal power plants*. GB 13223-2003. Beijing, China: State Environmental Protection Administration.

Taylor, M. R., Rubin, E. S. & Hounshell, D. A. 2005. Control of SO_2 emissions from power plants: A case of induced technological innovation in the U.S. *Technological Forecasting and Social Change*, 72, 697–718.

Tietenberg, T. H. 1990. Economic instruments for environmental regulation. *Oxford Review of Economic Policy*, 6, 17–33.

The U.S. Congress. 1990. *Clean air act amendments 1990*. Washington, DC: The U.S. Congress.

Xu, Y. 2011. China's functioning market for sulfur dioxide scrubbing technologies. *Environmental Science and Technology*, 45, 9161–9167.

Xu, Y., Williams, R. H. & Socolow, R. H. 2009. China's rapid deployment of SO_2 scrubbers. *Energy & Environmental Science*, 459–465.

Xu, Y., Yang, C. J. & Xuan, X. W. 2013. Engineering and optimization approaches to enhance the thermal efficiency of coal electricity generation in China. *Energy Policy*, 60, 356–363.

Yan, G., Yang, J., Wang, J., Chen, X. & Xu, Y. 2009. Carry out SO_2 emission trading program to set up long-term mechanism for emission reduction. *In:* Wang, J., Lu, J., Jintian, Y. & Li, Y. (eds.) *Environmental policy research series*. Beijing, China: China Environmental Science Press.

Zhang, B., Fei, H., He, P., Xu, Y., Dong, Z. & Young, O. 2016. The indecisive role of the market in China's SO2 and COD emissions trading. *Environmental Politics*, 25, 875–898.

Zhang, B., Zhang, H., Liu, B. B. & Bi, J. 2013. Policy interactions and underperforming emission trading markets in China. *Environmental Science & Technology*, 47, 7077–7084.

Zhao, Y., Wang, S., Duan, L., Lei, Y., Cao, P. & Hao, J. 2008. Primary air pollutant emissions of coal-fired power plants in China: Current status and future prediction. *Atmospheric Environment*, 42, 8442–8452.

6 Policy implementation[1]

1 Goal-centered policy implementation

In a country with effective rule of law, law enactment and policy making are the most important step for environmental protection, while implementation is more or less expected, although some bumps may still exist. China does not have sound rule of law, and thus, its policy implementation could be even more important than policy making. In the United States, after the Acid Rain Program in the Clean Air Act Amendments (1990) was enacted, law enforcement was largely the responsibility of the administrative branch. The rule of law obliges the administration to enforce the law. However, in China, no such tradition has been established to ensure that laws and policies will be genuinely implemented. Furthermore, China's policy implementation is heavily decentralized to local governments (Chapter 3), while the U.S. federal government has a relatively much stronger capacity for implementing their own policies. A key difference between China and the United States is that China should first mobilize its decentralized policy implementers before witnessing significant efforts and sulfur dioxide (SO_2) mitigation. China relies on the goal system to mobilize ministries at the central government and, more important, local governments for policy making and implementation, as discussed in Chapter 4.

Environmental compliance in China was indeed weak but has been improving steadily. China has made much progress in the past 15 years to reverse the earlier poor implementation of environmental policies (Jin et al., 2016). Coal-fired power plants in China have nearly universally installed SO_2 scrubbers, already 94.4% as of 2013 (Figure 5.11). Although most SO_2 scrubbers in China today do operate properly to greatly contribute to the deep reduction of SO_2 emissions (Figure 1.8), evidence of their misreporting and cheating was widely present to indicate serious noncompliance problems. A study showed that many factories in China were primarily concerned about minimizing operation costs and only operated their pollutant removal facilities when an inspection was imminent (OECD, 2006). Official data reported that SO_2 emissions from the power sector in 2007 were 11.5 million tons (Ministry of Environmental Protection, 2006–2009), but an independent study estimated that 16.4 million tons were emitted in that year (Lu et al., 2010). In addition, official data announced that in 2007, 73.2% of SO_2

was removed from coal-fired power plants that had SO_2 scrubbers (Ministry of Environmental Protection, 2009b). In Jiangsu Province, which had a relatively good track record on environmental protection, however, the rate was found to be only about one third in the first few months of 2007 (SERC, 2009). Especially before June 2007, cheating was widespread (Figure 6.1). Although almost all coal-fired power plants generally reported that their SO_2 scrubbers were operating normally, later confirmed data found the operating time to be much shorter (Figure 6.1). For those in operation, their SO_2 removal efficiencies were often much lower than required (SERC, 2009). However, after July 2007, a great majority of their SO_2 scrubbers were operating for more than 90% of the time and were achieving SO_2 removal efficiencies of over 90% (Jiangsu Department of Environmental Protection, 2007–2009). Data from the Ministry of Environmental Protection reported that SO_2 scrubbers had already been removing 78.7% of SO_2 from associated coal-fired power plants in 2008 (Ministry of Environmental Protection, 2009b), indicating that they were largely operating as they were supposed to do. This chapter evaluates such transition and examines how the compliance decisions were reversed.

After SO_2 scrubbers are installed, the managers of coal-fired power plants decide whether to operate them or not. The willingness to install SO_2 scrubbers does not

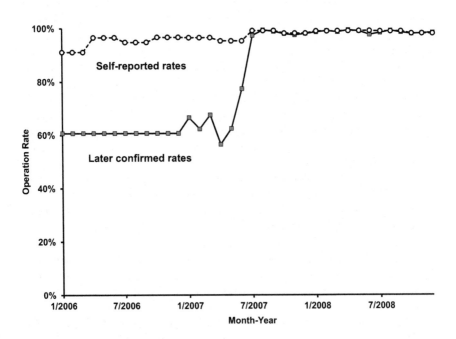

Figure 6.1 The operation of SO_2 scrubbers in Jiangsu Province, including self-reported operation rates and later confirmed operation rates

Source: The Economic & Trade Commission of Jiangsu Province, 2009; Jiangsu Department of Environmental Protection, 2007–2009; State Electricity Regulation Commission (Nanjing office), 2009.

necessarily mean that the incentives are strong enough for their proper operation. In addition, China's SO_2 scrubbers vary greatly in sizes, technology types, sulfur contents, costs of reagents and local environmental governance effectiveness. A significant variance should exist in their operation especially across provinces.

Studies have shown that three conditions are favorable to ensure compliance with environmental legislation: low compliance costs, high penalties for noncompliance and a high probability of catching noncompliance (Cohen, 1999; Helland, 1998; Becker, 1968). The latter two conditions are complementary to each other. In 2000, Blackman and Harrington judged that in China, "both the probability of getting caught for underreporting and the penalty for doing so are quite low" (Blackman and Harrington, 2000). A 2009 IEA (International Energy Agency) report claimed that the operation of China's SO_2 scrubbers was problematic due to high operation costs and ineffective environmental regulation (IEA, 2009). As the crucial factor to decide the probability of catching noncompliance, the importance of an effective compliance monitoring system has been widely recognized for implementing environmental policies and achieving their intended objectives (Lu et al., 2006; Raufer and Li, 2009). Monitoring and site inspection are essential for catching offenders but are subject to the constraints of high costs and limited budgets (Arguedas, 2008). The problem is especially serious in developing countries (McAllister et al., 2010; Blackman and Harrington, 2000). To enforce the SO_2 allowance trading scheme in the United States, the strategy was to install continuous emissions monitoring systems (CEMSs) with "periodic quality control tests of monitoring devices to maintain the accuracy of emissions data" (The U.S. Congress, 1990; Stranlund and Chavez, 2000). Large polluters can attract more attention. A study of the U.S. steel industry found that large polluting plants attracted more scrutiny than their smaller counterparts, regardless of how good their compliance record was (Gray and Deily, 1996).

Similar to policy making, policy implementation in China is also goal-centered, under which actions from the central and local governments focus more on whether they can contribute to goal attainment and less on whether policies are genuinely implemented. Heavily decentralized policy implementation facilitates their selective and goal-centered enforcement efforts. Such selective policy implementation also indicates that many rules in China are not followed or respected even by governments, which results in the weak rule of law. Given the political, economic, social and technological feasibility of implementation, as well as probable capacity constraints, those policies that can lead to more pollution mitigation have higher probabilities of being prioritized in implementation. This will trigger the policy evolution through implementation selection, as discussed in Chapter 5. For implementing a given policy in China's context of originally low environmental compliance rates, those factors that contribute to their enhancement are strengthened selectively and sequentially, depending on how progress can be made more effectively and efficiently with corresponding efforts. In a certain period, compliance costs are determined by technological statuses and market conditions, which are largely not decided directly by the environmental administration. In the longer term, technologies may evolve and costs may go down, as examined in detail in

Chapter 7. The other two factors, penalties for noncompliance and environmental compliance monitoring, are primarily examined in this chapter.

This goal-centered policy implementation echoes the comparative advantage theory that was originated by David Ricardo to analyze the development of international trade (Ricardo, 1817). In a two-country, two-product model, even though a country may have lower productivity or *absolute disadvantage* in producing both products, it still could specialize in and export one product based on *comparative advantage* and import only the other. Heckscher and Ohlin further developed the model to attribute the origin of comparative advantage in a country's factor endowment (Ohlin, 1967). It later became the foundation of a development theory that argued that a country should base its development on its comparative advantage (Chenery, 1961). Lin et al. employed this theory to explain the rapid economic growth of China and other countries (Lin et al., 2003). Before the economic reform in 1978, China adopted a leap-forward strategy to develop capital-intensive heavy industries against its comparative advantage, and this resulted in slow and unsustainable economic growth, whereas after the reform, the comparative advantage of labor was better utilized to achieve rapid economic growth and upgrading (Lin et al., 2003).

From a status of prevalent noncompliance, goal-centered policy implementation suggests making progress according to the contingent comparative advantage of alternative paths for achieving goals. When policy implementers decide which path could better serve the SO_2 mitigation goals, the chosen path should follow the direction whereby the effort's "productivity" is comparatively higher. In other words, easier measures are taken first before moving to more difficult measures, although many problems exist in the process.

This chapter examines the progress of three key measures. Penalties for noncompliance were first increased. A 2007 policy provided subsidies to coal-fired power plants for normally operating their SO_2 scrubbers, but nonoperation would incur a penalty of five times. Managers of those coal-fired power plants, mostly state-owned, would lose their jobs if cheating were caught. These penalties were relatively easier to be made available, while the more difficult environmental compliance monitoring was strengthened in following two steps to enhance the probability of catching noncompliance. First, more resources were made available to support the conventional environmental compliance monitoring system that features monitoring, reporting and verification (MRV). CEMSs also played a key role to signal potential noncompliance. The strategy worked well for coal-fired power plants that tend to be large, making frequent inspections little constrained by the shortage of inspectors. Collusion was also made more difficult for improving data quality. Furthermore, new technologies for environmental compliance monitoring are rapidly emerging and evolving, including sensors, satellites and social media. They tend to be much cheaper in monitoring one polluter but less accurate for legally confirming compliance statuses and issuing penalties, while conventional technologies are much more expensive but, if working smoothly, can meet the legal accuracy requirements. In the 2010s and, especially, since 2015, the Chinese government has been actively developing and integrating these big data technologies into governance. In environmental protection with millions of polluting sources scattered across China's vast geographic landscape, environmental compliance monitoring is one primary field to apply these new technologies for achieving higher compliance rates without demanding more resources.

2 Compliance on the operation of SO$_2$ scrubbers

2.1 SO$_2$ scrubber technologies

Compliance costs of SO$_2$ scrubbers are mainly for their operation as well as maintenance. A comprehension of SO$_2$ scrubber technologies is accordingly essential to understand how coal-fired power plants may cheat on compliance and how the government could catch such noncompliance.

SO$_2$ scrubbers (or flue gas desulfurization) have various technology types. Wet scrubbers are the most applied technology with SO$_2$ removal efficiencies normally over 90% (Figure 5.15). This section introduces major features associated with wet scrubbers and briefly compares these with dry scrubbers. After the flue gas comes out of a dust-removal facility (generally electrostatic precipitator, or ESP, in China's coal-fired power plants), it will be directed to an SO$_2$ scrubber system. The first step is often to pass the flue gas through fans or boosters, which facilitate the flow and adjust the velocity of the flue gas to be in a desirable range for the best performance of the SO$_2$ scrubber. Then the flue gas enters an absorber tower, where actual SO$_2$ removal happens. Generally speaking, coal-fired power generation units of 300 MW or over should have their own absorber towers and two units of 200 MW or less could share one (NDRC, 2004). The flue gas enters the absorber tower at the lower-middle part and moves upward. Limestone slurry mixed with products of chemical reactions fills the lower part of the absorber tower and is lifted by several circulation pumps to the upper part. Special nozzles are used to spray the slurry for the maximization of SO$_2$ removal efficiency. The falling slurry droplets contact the flue gas physically and remove approximately 90% to 95% of SO$_2$ through chemical reactions. Simply put, the main and overall reaction is $CaCO_3 + SO_2 + H_2O \rightarrow CaSO_3 + CO_2 + H_2O$. Air is blown into the slurry pool at the lower part of the absorber tower to force the oxidization of SO_3^{2-} and make gypsum $(CaSO_4 2H_2O)$: $CaSO_3 + \frac{1}{2}O_2 + 2H_2O \rightarrow CaSO_4 2H_2O \downarrow$.

Before the flue gas exits from the top of the absorber tower, it passes through equipment that removes mist. Because the processing removes much heat from the flue gas and the efficient outflow from a chimney (often over 200 m high for coal-fired power plants) requires the flue gas to be above a certain temperature, a gas–gas heat exchanger may be included to heat the outlet flow gas with the inlet flue gas to raise its temperature.

Two important side systems are respectively for the preparation of limestone slurry and the production of gypsum. Limestone is crushed and mixed with water to make limestone slurry. Fresh limestone slurry enters the absorber tower often through circulation pumps. The bottom slurry with gypsum is pumped out and filtered to separate gypsum. The wastewater is sent to a treatment system.

SO_2 removal efficiency can be controlled by adjusting various factors, including the contact time length between the flue gas and the limestone slurry droplets, calcium-to-sulfur ratio (or Ca/S ratio) and liquid-to-gas ratio (or L/G ratio). The contact time depends on the velocity of the flue gas and the path length before its leaving the point where limestone slurry is injected. The height of the absorber tower is an influential factor determining the path length. Another factor is associated with the different injection heights of circulation pumps. Higher injection points indicate a longer path for contact and reaction. When the electricity generation unit is not in full load with less flue gas, not all circulation pumps will have to be operated. Then the choice of different circulation pumps could make some difference in the SO_2 removal efficiency. However, higher absorber tower and longer contact path correspond to higher electricity consumption to lift limestone slurry.

In the L/G ratio, the liquid refers to the volume of limestone slurry dropping from the upper part of the absorber tower, or circulated liquids. The gas is the volume of flue gas entering the absorber tower. Higher L/G ratio leads to higher SO_2 removal efficiency because the chance is higher for an SO_2 molecule to be absorbed. It is controlled through circulation pumps: if the flue gas volume does not change, turning on more pumps indicates a higher L/G ratio. Since the number and power of circulation pumps are fixed after an SO_2 scrubber comes online, the liquid volume has an upper limit, which restrains the maximum contribution of enhancing L/G ratio to increase SO_2 removal efficiency.

Ca/S ratio is the molar ratio between calcium carbonate ($CaCO_3$) and sulfur oxides (SO_x, dominantly SO_2). In a perfect situation, as predicted in the chemical reaction introduced earlier, the ideal Ca/S ratio is 1 to remove all SO_2 and use up all limestone. But in the actual situation, not all limestone will be consumed and not all SO_2 will be removed. SO_2 wet scrubbers can achieve high efficiencies in both aspects. As a result, the actual Ca/S ratio is only a little higher than 1, usually around 1.03 for China's wet scrubbers (Wu and Qian, 2007). Because the inlet quantity of SO_2 changes with the volume of flue gas and the SO_2 concentration, even though the Ca/S ratio is maintained stable, the rate of adding limestone to the system will still change. On the other hand, the workload of removing SO_2 could become too heavy when the actual sulfur content exceeds the designed level by a significant margin. In this situation, when all circulation pumps are turned on and the L/G ratio has reached its maximum, the only major method to maintain a required high SO_2 removal efficiency is to enhance the Ca/S ratio. However, a much higher Ca/S ratio than the designed level will not only add costs but also will more likely clog the system and barricade its normal function. The adjustment of the Ca/S ratio is through controlling the pH value of the limestone slurry in the absorber tower. In daily operation, the pH value should be maintained within a range. A pH value above the normal range indicates excessive limestone and that the injection rate of fresh limestone slurry should be reduced.

SO_2 wet scrubbers consume about 1% of the electricity generated from the corresponding power generation units. The rate could be as high as 3.5% when high-sulfur coal is burned with a heavy workload of SO_2 removal. China's coal-fired power plants consumed, on average, 6.79% of the electricity they generated in

2008 (SERC et al., 2009), in which SO_2 scrubbers accounted for a notable share. Significant electricity-consuming components of SO_2 scrubbers include the fans, circulation pumps and limestone slurry preparation system.

Two major economies of scale are associated with SO_2 scrubbers in construction. First, the size of an absorber tower is largely determined by the volume of flue gas or the scale of the corresponding power generation unit. A larger volume of flue gas or a larger scale in megawatts leads to lower average costs for each unit of flue gas treated or each megawatt. Because absorber towers are responsible for a large part of the capital costs, this economy of scale could significantly reduce the unit capital costs. Since most nonpower SO_2 emission sources consume much less coal and do not generate large enough volume of flue gas to provide the economy of scale, the unit costs of SO_2 scrubbers are often more expensive.

Second, higher SO_2 concentration in the inlet flue gas, or higher sulfur input rate, raises capital costs for each unit of flue gas treated because they require larger systems of limestone preparation and gypsum handling, as well as probably higher absorber tower and more circulation pumps. Economy of scale can also be realized for these systems to treat each unit of SO_2. SO_2 concentration in the inlet flue gas is mostly determined by two factors: sulfur and thermal contents of coal. The link with sulfur contents is quite straightforward: if different types of coal only differ in sulfur contents, higher sulfur contents indicate more SO_2 in a roughly equal amount of flue gas. With the same thermal efficiency, the volume of flue gas mainly depends on the thermal input. Then lower thermal contents of coal mean that more coal has to be burned for the required thermal input, and accordingly, more SO_2 will be generated. Accordingly, sulfur content per unit of energy is a better indicator of SO_2 concentration in the inlet flue gas of SO_2 scrubbers.

The product of SO_2 scrubbers is gypsum. Depending partly on the quality, it can either be sold in the market or go to landfill. A significant market for gypsum is in building materials.

The operation and maintenance (O&M) of SO_2 scrubbers are associated with costs in materials (including mainly limestone, electricity and water), labor and maintenance. The sale of gypsum could earn some revenue but often only at an insignificant portion. If the quality of SO_2 scrubbers remains about the same, maintenance costs are positively related to the capital investment of SO_2 scrubbers. As a result, larger scales of SO_2 scrubbers are linked with lower maintenance costs on the bases of megawatt-hour or ton SO_2 removed. Similarly, economy of scale is also relevant to labor costs, but the impact on overall O&M costs is constrained by the insignificant share of labor costs; for example, my field trip to China's coal-fired power plants found that roughly 15 workers were required to run the SO_2 scrubber and ESP for a 300-MW plant in 2008. Their total annual costs could be about 1 million RMB. The average O&M costs of China's SO_2 scrubbers were about 15 RMB/MWh, indicating that the total O&M costs would be approximately 23 million RMB if the capacity factor was about 5,000 hours/ year. Then the share of labor costs was less than 5%.

Materials comprise most of operation costs. In normal operation, the Ca/S ratio remains fairly stable, and thus, the limestone consumption is about linearly related to sulfur input. Electricity consumption for running SO_2 scrubbers can be roughly divided into three major parts: in fans that are mainly associated with the flue gas volume, in the handling of limestone and gypsum that is affected by sulfur input and in circulation pumps connected with both. Most water consumption is in the form of evaporation to the flue gas in the absorber tower, and the water is released to the atmosphere together with the cleaned flue gas. As a result, water consumption is mainly correlated with the volume of flue gas and is also affected by economy of scale.

Because of its wide availability and low costs, limestone is the dominant absorbing reagent in wet scrubbers. However, other alkaline reagents are sometimes applied, such as seawater and alkaline wastewater. Furthermore, SO_2 scrubbers can be dry. Lime (CaO) is often used as the absorbing reagent. Because of its much lower utilization rate, the Ca/S ratio has to be much higher (e.g., 1.3~1.5). Generally, the SO_2 removal efficiency is in the range of about 70% to 80%, lower than that of wet scrubbers. The capital costs of dry scrubbers are lower, but the operation and maintenance costs are higher (EPA, 2003).

2.2 Noncompliance behaviors

The managers of coal-fired power plants have strong incentives to avoid the costs of O&M. Data from Jiangsu Province showed that from 2006 to June 2007, the self-reported operation rates (the percentage of time that an SO_2 scrubber is in operation alongside the corresponding power generation unit) from coal-fired power plants were significantly higher than the values that were later confirmed, likely through other relevant data such as limestone consumption, gypsum production and electricity consumption (respectively, more than 90% and about 60%; Figure 6.1). The discrepancy reflects the likely magnitude of misreporting. This section discusses several prominent problems that emerged from the author's interviews and the literature. These problems prevented the proper operation of SO_2 scrubbers and caused very significant uncertainty in estimating SO_2 emissions from coal-fired power plants.

Typical noncompliance behaviors could include the following: first, SO_2 emissions may be underreported and the quality of SO_2 scrubbers could be poor. Coal-fired power plants underreported SO_2 emissions to pay a lower effluent discharge fee and to be seen as complying with regulations. In 2007, 98% of coal consumed in China's power plants was raw coal (National Bureau of Statistics, 2008). Coal-fired power plants were allowed to pick out coal stones from received raw coal to calculate actual coal consumption. In interviews, the author found that coal stones were sometimes overreported. This factor could have led to a 1% to 2% underestimation of SO_2 emissions. Furthermore, China's coal-fired power plants usually had to use different coals with sulfur contents that could vary significantly. The instability of coal supply was confirmed by Steinfeld et al. (2009). It made the underreporting of sulfur contents harder to detect. Interviews in China's SO_2 scrubber companies found that many early scrubbers (e.g., before 2005) had serious quality problems. In order to reach designed SO_2 removal efficiencies, besides the replacement of malfunctioning equipment, a few SO_2 scrubbers even had to have their very expensive absorber towers retrofitted. The main reason for the faults in the SO_2 scrubbers was that they were designed on the basis of underreported sulfur contents. In China's first public and high-profile statement to penalize the abnormal operation of SO_2 scrubbers, instability and bad quality were particularly pointed out, and three power plants were found to use coal with much higher sulfur than the designed levels (Ministry of Environmental Protection, 2008). At the design stage of SO_2 scrubbers, if the managers of coal-fired power plants had been underreporting sulfur contents in the past, they would continue to do so to conceal their guilt. Some managers did not plan to operate their SO_2 scrubbers initially and were not concerned about their quality. They installed the SO_2 scrubbers purely in order to comply with the government's requirements and to qualify for a subsidy for generating desulfurized electricity. The managers wanted to minimize capital costs through underreporting sulfur contents. When inspections were known in advance, reaching the required SO_2 removal efficiencies was not a problem because coal-fired power plants often kept some low-sulfur coal in reserve on-site.

Second, illegal bypass ducts may be used to avoid flue gas treatment. Many SO_2 scrubbers had bypass ducts to allow the flue gas to exit without going through the SO_2 scrubber systems. The purpose was to enable electricity generation when SO_2 scrubbers had minor problems and need to be shut down temporarily. In a 2007 policy, coal-fired power plants were not penalized provided that their SO_2 scrubbers were properly operating for at least 90% of the time (NDRC and SEPA, 2007b). However, bypass ducts also provided opportunities to avoid the operation of SO_2 scrubbers when they functioned normally. Six coal-fired power plants were penalized for illegally using bypass ducts and leaving some flue gas untreated in 2007 and 2008 (Ministry of Environmental Protection, 2008, 2009c).

Third, data from CEMSs may also be inaccurate and manipulated. CEMSs could greatly enhance environmental monitoring capacity. China had 60 SO_2 scrubbers at the end of 2004 (Ministry of Environmental Protection, 2010a), while a general survey in 2004 found that about 400 CEMSs had been installed in 180 coal-fired power plants (Pan et al., 2005). The author's site visits and Steinfeld et al. also found that CEMSs were being widely used (Steinfeld et al., 2009). As far as cost was concerned, there was little reason for coal-fired power plants to resist the installation of CEMSs. Two CEMSs in Plant 3 in Table 6.1 cost about US$132,000, only 0.5% of the capital costs of the plant's SO_2 scrubbers. However, CEMSs may not report credible and reliable data. One concern was over the quality of the equipment used. CEMSs cost much more in the United States: according to a cost model from the U.S. Environmental Protection Agency (EPA), it generally required more than half a million dollars for one set (The U.S. EPA, 2007). The 2004 general survey found that only 20% of the CEMSs in China were functioning normally (Pan et al., 2005), local environmental protection bureaus generally refused to accept data from CEMSs and only one was recognized as a credible data source for the purposes of levying the SO_2 effluent discharge fee (Pan et al., 2005). Later on, CEMSs were officially accepted as data sources after

Table 6.1 Data on SO$_2$ scrubbers in China's seven coal-fired power plants

Plant No.	Region	New or Retrofit	SO$_2$ scrubber type	Sulfur content	Operation & maintenance (O&M) costs	Profit margin of electricity generation	Price premium for desulfurized electricity	Effluent discharge fee **
				%	US$/MWh	US$/MWh	US$/MWh	US$/MWh
Plant 1	Southwest	New	Wet	3.0%	3.7	~14.6	2.2	2.0
Plant 2	Southwest	Retrofit	Wet	4.0%	~4.1	>0		2.8
Plant 3	Southwest	Retrofit	Wet	3.5%			2.2	2.5
Plant 4	East	Retrofit	Wet	1.0%*	1.8		2.2	0.7
Plant 5	East	New	Wet	0.5%	2.2	>>14.6	2.2	0.3
Plant 6	East	Retrofit	Wet	1.0%	<2.2	<7.3	2.2	0.7
Plant 7	East	Retrofit	(1): dry; (2): wet	0.7~1.1%	(1): >3.7; (2): <3.7		3.7	0.4–0.7

Note: The data were collected in the author's interviews in June and July 2009. The costs and profit margin reflect their most recent estimation generally for 2008. The original currency units were in Chinese RMB. In the conversion to U.S. dollars, the exchange rate on December 31, 2008, is used: US$1 = 6.83 RMB. Upon the request of the interviewees, the names of coal-fired power plants are intentionally not shown.

* No clear information on sulfur contents was disclosed in Plant 4. Comparing with other plants in the eastern provinces, 1.0% is used here for later analysis. ** The effluent discharge fee is calculated to reflect the additional payment if shutting down SO$_2$ scrubbers. The fee rate refers to the level of US$0.092/kg SO$_2$ (State Development Planning Commission et al., 2003). The intermediate assumptions are sulfur retention rates in ash, 20%, as assumed in compiling China's official data (State Council, 2007b); SO$_2$ removal efficiencies of wet scrubbers, 95%, according to the author's interviews; thermal efficiencies of electricity generation; and the standard levels at the corresponding unit scales (Zhejiang Bureau of Quality and Technical Supervision, 2007).

their online connection with provincial environmental protection bureaus. However, the author's interviewees still said that they did not fully trust data from CEMSs. The locations of the sensors could affect the readings of CEMSs, and data reporting could also be manipulated. In 2008, three coal-fired power plants were caught illegally setting up ceilings of outlet SO_2 concentrations that could be reported (Ministry of Environmental Protection, 2009c).

Fourth, coal-fired power plants could either cheat or even collude with environmental compliance inspectors. Data from CEMSs were compared quarterly with direct measurements to verify accuracy (State Council, 2007b). A survey in China found that multiple inspections per annum tended to deter violation, no matter which government level inspectors were from (Lu et al., 2006). However, the effectiveness of site inspections could be constrained. According to the author's interviews, many plants were able to raise the removal efficiency of their SO_2 scrubbers from zero to the designed level in half an hour and significantly more quickly if from an intermediate level. Some plants therefore kept their scrubbers either turned off or on low power and put them in full operation only when an inspection was imminent, enabling them to keep costs down while also passing the inspection. Even if abnormal operation were caught, a solution could be to collude with inspectors through bribes.

3 Reversing noncompliance: penalty

The proper operation of SO_2 scrubbers demands strong enough incentives to overcome the hurdle of the high O&M costs. In the United States, the average O&M costs in 2008 were US$1.55/MWh (EPA and DOE, 2010). These figures could hardly be extrapolated for China because of the great differences in the capital costs of SO_2 scrubbers, labor costs and other items. My interviews collected relevant data from six coal-fired power plants, as presented in Table 6.1. The O&M costs varied from US$1.8 to 4.1/MWh, all above the average in the United States. Sulfur contents were the most influential factor: Plants 1 and 3 burned coals with approximately 3% to 4% sulfur content, and their O&M costs were roughly twice as high as those in Plants 4, 5, 6 and 7, which burned coals with 1% sulfur content or less. The O&M costs could be used as the average marginal costs of operating SO_2 scrubbers. In generating electricity, several coal-fired power plants that the author visited had gross profit margins from not much above zero to significantly over US$14.4/MWh (including the O&M costs of SO_2 scrubbers and the price premium for desulfurized electricity).

A survey of China's inspection authorities and polluting firms found that fines for noncompliance were often not high enough to deter potential offenders (Lu et al., 2006). The initial SO_2 effluent discharge fee was about 0.20 RMB/kg (US$0.031/kg) in most provinces and lower than the marginal abatement costs in China's large plants (Cao et al., 1999; Dasgupta et al., 1997). Polluting firms may simply pay to pollute. When facing a penalty, the first reaction of polluting firms was to negotiate with environmental protection bureaus or ask for the interference of local governments (Lu et al., 2006), which compromised the penalty.

In July 2005, China's SO_2 effluent discharge fee was raised from US$0.031/kg in 2003 to US$0.092/kg (State Development Planning Commission et al., 2003). However, as converted to US$/MWh in Table 6.1, it was still too low to overcome the hurdle of the much higher O&M costs. Another increase was scheduled in 2007 to reach US$0.18/kg in three years (State Council, 2007a), but the exact schedule varied from province to province. In Jiangsu Province, the higher rate had been in effect since July 2007 (Jiangsu Department of Environmental Protection, 2008), but in Henan Province, the lower rate was still being applied in the first quarter of 2010 (Henan Department of Environmental Protection, 2010). With the higher rate, coal-fired power plants burning high-sulfur coals would find operating SO_2 scrubbers cheaper than paying the effluent discharge fee (Plants 1 and 3 in Table 6.1). However, for others (Plants 4, 5, 6 and 7) when facing only this policy, the rational decision was to pay the fee.

Another policy was introduced in 2004. If new coal-fired power plants came online together with SO_2 scrubbers, the desulfurized electricity could enjoy a price premium of US$2.2/MWh (NDRC and SEPA, 2007a). In June 2006, the policy extended to cover all SO_2 scrubbers, including retrofitted ones (NDRC and SEPA, 2007a). Some coal-fired power plants were awarded higher price premiums, such as Plant 7 in Table 6.1. The price premium and the effluent discharge fee together were a little higher than the O&M costs (Table 6.1), but the small difference indicated that the proper operation would be a rational decision only when most nonoperation cases were caught.

The 11th Five-Year Plan witnessed sharp increases in noncompliance penalties. In 2007, a harsh penalty measure was associated with the price premium for the first time. If the operation rate of an SO_2 scrubber were lower than 80%, a penalty of US$11.0/MWh would be issued for any additional non-desulfurized electricity generation (NDRC and SEPA, 2007b). The required minimum probability of catching nonoperation became much lower to induce the proper operation of SO_2 scrubbers. For Plant 3 in Table 6.1 burning high-sulfur coal, corresponding to the effluent discharge fee of US$0.092/kg SO_2, a risk-neutral manager would decide to operate SO_2 scrubbers properly if the probability of catching nonoperation exceeded 26% (see Table 6.2 for the calculation formula). For Plants 4 and 5 burning low- to medium-sulfur coals, the minimum probability was about one seventh. Furthermore, additional penalties were introduced on the managers of coal-fired power plants. In China, almost all coal-fired power plants were owned by the state. The nonoperation of SO_2 scrubbers could increase profit and benefit the managers' career and salary. However, according to formal regulations (NDRC and SEPA, 2007b) and the author's interviews, cheating and nonoperation could lead to the removal of the managers. They had to calculate the risk for themselves.

The penalties in 2007 also aimed for minimizing potential moral hazard when SO_2 scrubbers occasionally had to stop operating due to accidents, malfunctions or other reasons. While they were out of action, SO_2 emissions could be controlled either by minimizing the sulfur content of coal or by shutting down electricity generation. China would issue no penalty as long as the operation rate were above 90%, a mild penalty of US$2.2/MWh if the rate were between 80% and 90% and

Table 6.2 Decision scenarios for the managers of coal-fired power plants

Scenario	SO_2 scrubbers functioning	SO_2 scrubbers operating	Electricity generation	Net revenue of a coal-fired power plant
(1)	Yes	Yes	Yes	Profit margin
(2)	Yes	No	Yes	Profit margin + O&M costs – C% × (Price premium + Discharge fee + Penalty)
(3)	No	No	Yes	Profit margin + O&M costs – C% × (Price premium + Discharge fee + Penalty)
(4)	No	No	No	0

Note: C% is the actual probability of catching the nonoperation of SO_2 scrubbers. The proper operation of SO_2 scrubbers, when they function, requires that the net revenue in scenario (1) is greater than that in scenario (2). The corresponding condition can be calculated as $C\% > \dfrac{\text{O\&M costs}}{\text{Discharge fee} + \text{Price premium} + \text{Penalty}}$. When SO_2 scrubbers do not function, the discontinuation of electricity generation becomes a rational decision when the net revenue in scenario (4) is greater than that in scenario (3), or $C\% > \dfrac{\text{Profit margin} + \text{O\&M costs}}{\text{Discharge fee} + \text{Price premium} + \text{Penalty}}$. Because profit margins are generally positive, it is accordingly easier to push for the proper operation of SO_2 scrubbers when they function than to ask coal-fired power plants to discontinue electricity generation when they do not. In order to encourage coal-fired power plants to fix malfunctioning SO_2 scrubbers as soon as possible, the rational decision when SO_2 scrubbers function should generate greater net revenue than that with malfunctioning SO_2 scrubbers. The condition is fulfilled when $C\% > \dfrac{\text{O\&M costs}}{\text{Discharge fee} + \text{Price premium} + \text{Penalty}}$.

a harsh penalty of US$11.0/MWh if the rate were under 80% (NDRC and SEPA, 2007b). Because it was expensive to restart electricity generation, the O&M costs of SO_2 scrubbers may not be critical in the decision making when SO_2 scrubbers could get fixed soon.

When problems have to take much time to fix – for example, several weeks – and the penalty of US$11.0/MWh is applied, the economic incentives should make it a rational decision to discontinue electricity generation for many coal-fired power plant managers. Electricity generation without operating SO_2 scrubbers earned a profit margin and avoided the O&M costs of SO_2 scrubbers, but if the non-operation of SO_2 scrubbers were caught, coal-fired power plants would need to return the price premium and pay the effluent discharge fee as well as the penalty. Many coal-fired power plants might continue generating electricity as long as the probability of catching the nonoperation of SO_2 scrubbers was low enough (see Table 6.2 for the specific calculation). For coal-fired power plants with large profit margins (such as Plant 5 in Table 6.1), electricity generation should continue even when nonoperation could not be hidden at all. However, when the author visited Plant 5, electricity generation in one system had been discontinued for several weeks due to its malfunctioning SO_2 scrubber. Personal penalties on the managers could have played a role. Furthermore, even if the decision was to continue electricity generation, a high-enough probability of detection was still necessary to encourage coal-fired power plants to fix malfunctioning SO_2 scrubbers as soon as possible (see Table 6.2 for the specific calculation). If the actual probability was not expected to reach this level, there would be little concern about the quality of SO_2 scrubbers, as in the early years.

Furthermore, coal-fired power plants should also comply with regulations on SO_2 removal efficiency and effluent emission standards (NDRC and SEPA, 2007b; SEPA and General Administration of Quality Supervision Inspection and Quarantine, 2003; MEP and AQSIQ, 2011). Technically in practice, a coal-fired power plant could choose a designated SO_2 removal efficiency. For example, higher ratios of Ca/S (the molar ratio between $CaCO_3$ and SO_x) or L/G (the liquid-to-gas ratio in volume) would remove more SO_2 from the flue gas. Reasonably, if not regulated, a coal-fired power plant could lower SO_2 removal efficiencies to reduce costs. On the other hand, because of changing sulfur contents and workload, SO_2 concentration and flue gas volume were not stable. Scrubbers' capability to track the changes – with the same methods of adjusting SO_2 removal efficiencies – was necessary for their reliable operation.

The Chinese central government mandated minimum SO_2 removal efficiencies being established (NDRC and SEPA, 2007b) and provincial governments were in charge of the details. For example, when SO_2 removal efficiencies were lower than predetermined levels (generally 90% for wet scrubbers), Henan Province simply counted the time as nonoperation (Henan Development and Reform Commission and Henan Environmental Protection Bureau, 2007). In normal conditions, the incentives were strong enough to make SO_2 scrubbers reach the required levels of SO_2 removal efficiencies. Two actual cases from the author's field trip could demonstrate the decisions. In the first case, sulfur contents went up significantly but were expected to be a temporary situation. The designed sulfur content for Plant 6's SO_2 scrubber was 0.84%, but for a period in 2008 when coal supply was constrained, the actual sulfur content was higher than 2%. Such a dramatic increase in sulfur content became a serious burden. To maintain SO_2 removal efficiencies over 90%, the solution was to raise the Ca/S ratio from the designed level of 1.03 to 1.3. In the second case, when the increased sulfur contents were expected to be long-lasting, a temporary solution would not be sustainable. One of the eight coal-fired power plants the author visited had to shut down and modify the original SO_2 scrubber to handle the much higher sulfur input rate. Particularly, the absorber tower became significantly taller by adding another section on the top of the original one. The pathway was accordingly longer for the flue gas and limestone slurry to contact and react. Additional circulation pumps could also be added to enhance the L/G ratio.

In order to better implement the incentives, responsible government agencies are specified: electric grid corporations were in charge of paying the price premium in time; provincial environmental protection bureaus collected effluent discharge fees; provincial price agencies were responsible to recover unjustified price premium according to actual operation rates (NDRC and SEPA, 2007b). Seven coal-fired power plants in 2008 and five in 2009 were penalized for cheating or nonoperation with the US$11.0/MWh penalty applied (Ministry of Environmental Protection, 2009c, 2008).

Central and local governments in China are not the only entities that have their tasks centered around goals. Because almost all coal-fired power plants in China were state-owned, they were also assigned quota or goals for their total SO_2 emissions (SEPA, 2006). Both goals and policies play crucial roles in their compliance decisions on the operation of their SO_2 scrubbers. In addition to financial penalties, administrative penalties were also applied for noncompliance. In environmental enforcement and compliance, decision makers at local governments, power corporations and coal-fired power plants also kept in mind their SO_2 emission caps or goals. If SO_2 removal efficiencies were too low and nonoperation was caught, the SO_2 emission permits could be used up soon. In addition, seriously abnormal operation of SO_2 scrubbers was publicly punished (MEP and NDRC, 2008; Ministry of Environmental Protection, 2009a), which could affect the career of the coal-fired power plants' managers. In the words of an interviewee, "it is not worthwhile for the managers of a coal-fired power plant to risk losing the positions to save money for the plant. Anyway, the money is not theirs, but the positions are."

4 Reversing noncompliance: environmental compliance monitoring

The effectiveness of environmental compliance monitoring determines the probability of catching noncompliance. China's emission data MRV system is largely bottom up, which could potentially suffer from two major challenges. The first challenge lies in the system's high costs. Compliance monitoring resource constraints exist in all countries, but the problem is especially daunting in developing countries, due to the high costs of compliance monitoring, limited resources, understaffed environmental agencies, inadequate training and technological support (Arguedas, 2008; McAllister et al., 2010; Blackman and Harrington, 2000; Russell and Vaughan, 2003; Pan et al., 2005). How to better utilize available resources is critical to determine the effectiveness of every domestic policy and international environmental treaty. Compliance monitoring is the most resource-consuming activity in enforcing environmental policies. For example, an emission trading scheme should effectively deter cheating and verify actual emission levels (Kruger and Egenhofer, 2006), while compliance monitoring was responsible for 69% of transaction costs for German companies in the European Union CO_2 Emission Trading Scheme (Heindl, 2012). The existence of many small and medium-sized polluters could seriously attenuate available resources, even in developed countries where the rule of law is generally well established. Due to the significant economy of scale, large point sources generally have lower compliance monitoring costs on a per-ton-emission basis and are often prioritized (Heindl, 2012; Gray and Deily, 1996). Because of China's sheer size, the large system involves many personnel and occupies substantial resources. In the 12th Five-Year Plan (2011–2015) alone, the Chinese government planned to invest 40 billion RMB (~US$5.9 billion) to enhance related environmental regulation capacity (MEP, 2013).

The second challenge is intentional data manipulation. Environmental monitoring and reporting in China generally must pass through, and be inspected by, polluting firms and various levels of local governments and relevant agencies before reaching the central government. Most environmental compliance capacities, such as personnel and governmental expenditure, are in local governments, while the central government is mainly in charge of policy making. Emissions of CO_2, SO_2 and NO_x are generally calculated via bottom-up energy consumption data and emission factors (Liu et al., 2015; Lu et al., 2011; Zhang et al., 2007). This approach is often subject to the influence of intentional distortions for the interest of stakeholders along the path (Tsinghua University, 2010). China has been exerting increasingly high pressure on local governments and energy-intensive firms to achieve top-down energy and emission control goals from the central government (Xu, 2011b). In comparison to the technologically challenging, economically expensive and politically difficult tasks of actual mitigation, it would be much more convenient to twist the reported numbers (Jin et al., 2016).

The objective resource constraint and the intentional data manipulation could seriously compromise data quality and thus the effectiveness of environmental compliance monitoring. Facing immense pressure of environmental crises, the Chinese government has been actively searching for potential solutions for enhancing environmental data quality.

4.1 Model construction

In order to understand China's environmental compliance monitoring in greater depth, a conceptual, computable model is constructed to simulate the evolution of compliance rates under different compliance monitoring strategies and how influential factors in three categories – pollution abatement costs, noncompliance penalty and, most important, compliance monitoring effectiveness – affect compliance decisions of polluters and thus the compliance rate. Mathematical details of the model are provided in the Appendix to this chapter.

This model stands on the shoulders of two pieces of research literature for creating a theoretical framework. The first well-developed economics literature of crime and punishment understands crimes as rational choices. Whether a polluter chooses compliance or noncompliance is based on comparing related costs and benefits (Polinsky and Shavell, 2000; Becker, 1968; Glaeser, 1999; Xu, 2011a; Shimshack, 2014; Levitt, 2004). If a polluter pondered not complying with an environmental regulation, pollution abatement costs could be saved as its expected benefits. However, such behavior would incur expected costs, which is a product of (1) penalty on noncompliance and (2) the probability of being caught. Risk-neutral rational polluters would choose environmental noncompliance if the expected benefits were greater than the expected costs. The compliance or noncompliance decision is assumed to be deliberate but not at random. The second mature literature, or a series of related literature, such as on policing, pollution control and tax evasion, examines how to enhance the probability of catching noncompliance. Compliance monitoring could apply various strategies for enhancing the probability with a given amount of resources, although the effectiveness is mixed. Levitt (2004) found that policing strategies are of only minor significance, while the number of police may explain a large proportion of the crime rate change. For tax compliance, endogenous audit selection rules screen taxpayers for potential auditing, but the impacts on compliance are mixed (Konrad et al., 2017; Vossler and Gilpatric, 2018). In epidemiology, strategies are developed to promote public health and enhance the rate of finding sick patients at early stages among a population (Bonita et al., 2006). A population would be first screened, and those with positive results would have to go through another round of more careful diagnosing for confirming whether they were true or false positive.

These two pieces of literature are integrated together in this study to simulate environmental compliance decisions. Two environmental compliance monitoring systems are proposed and simulated, as illustrated in Figure 6.2. The conventional system that is based on monitoring, reporting and verification is simplified to require governmental compliance monitoring resources primarily for site inspection. Adopting the terminology in epidemiology, the model refers to these activities as *diagnosing*. If a polluter were caught as being noncompliant, a penalty would be issued. The new compliance monitoring system inserts an additional step before diagnosing to actively *screen* polluters into high-risk and low-risk groups, with higher and lower probabilities of being noncompliant, respectively. Diagnosing with higher costs follows with site inspections or other more accurate means to confirm noncompliance only in the high-risk group. For the convenience

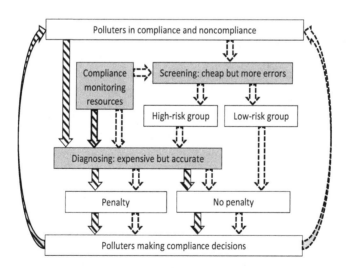

Figure 6.2 A conceptual model of environmental compliance monitoring

Note: The dash-line arrows indicate the screening system's flow, while the diagonal-pattern arrows refer to the diagnosing system's flow. Their major difference is the existence/absence of the screening step with screening technologies. The gray boxes show compliance monitoring resources that not only are allocated between screening and diagnosing technologies in the screening system but only to diagnosing technologies in the diagnosing system.

of discussion, the conventional system is referred to in this chapter as *the diagnosing system*, while the new system contains both screening and diagnosing, and it will be called *the screening system*. Numerous studies have applied the economic model of crime and punishment for understanding environmental noncompliance (Xu, 2011a; Shimshack, 2014; Guo et al., 2014). The compliance monitoring strategy with screening has also been widely applied in multiple fields (Konrad et al., 2017; Vossler and Gilpatric, 2018; Bonita et al., 2006).

4.2 Strengthening the conventional diagnosing system

China has made several prominent improvements in monitoring and site inspection to address the previously mentioned two challenges for enhancing the probability of catching the nonoperation of SO_2 scrubbers. First, more resources were made available for environmental compliance monitoring. The numbers of government employees at all levels increased from 46,984 in 2005 to 52,944 in 2009 and 61,668 in 2015 for environmental monitoring and from 50,040 in 2005 to 60,896 in 2009 and 66,379 in 2015 for inspection (Figure 3.1). Although still limited, the personnel resources had already been enough to have an intensive focus on SO_2 scrubbers in coal-fired power plants. Particularly, only 503 coal-fired power plants housed 461 GW SO_2 scrubbers (1,264 systems) at the end of 2009, and the largest 300 had a total capacity share of 82% (Ministry of Environmental Protection, 2010a). In 2013, 282 coal-fired power plants that were at or greater than 1 GW each had 470 GW SO_2 scrubbers in total, or 62.3% of all (Ministry of Environmental Protection, 2014). Government personnel are sufficient to follow these large plants closely and conduct inspections frequently.

The number of polluting sources (N) that require compliance monitoring varies dramatically, depending on focused polluter sizes, pollutants and other features. China conducted the first census of polluting sources with the census date being December 31, 2007, and pollution information for 2007, covering 5,925,576 polluting sources, including 1,575,504 industrial, 2,899,638 agricultural, 1,445,644 domestic and 4,790 centralized pollution control facilities (Ministry of Environmental Protection et al., 2010). In comparison, China's annual environmental statistics report focused on about one tenth of the polluting sources, being 161,598 industrial sources, 131,837 farms and 7,578 districts for animal husbandry and 6,910 water treatment plants, 2,315 municipal waste treatment facilities and 866 hazardous waste treatment facilities in 2015 (Ministry of Environmental Protection, 2002–2016). Among these sources, 68,121 polluting sources were under special supervisory monitoring (Ministry of Environmental Protection, 2002–2016).

Furthermore, in order to make the conventional diagnosing system more efficient, CEMSs have become critical to monitor the operation of SO_2 scrubbers especially since 2007 (NDRC and SEPA, 2007b). Six plants (all in Table 6.1 except Plant 2) allowed me to read the computer screens of their CEMSs. The values of SO_2 concentrations changed continuously, and different data were generally consistent. Many CEMSs and SO_2 scrubbers had been inspected once or twice a month. Because CEMSs transmitted data online and in real time, inspections often followed abnormal data reporting. Coal-fired power plants were informed in advance of some inspections, but in many other cases, inspections were unannounced. Inspectors had the right to enter coal-fired power plants without being delayed. In the plants that I visited, inspection vehicles generally needed just a few minutes to drive from the gates to the sites where the SO_2 scrubbers were installed. China was actively building up its site inspection capacity. The number of government inspectors at all levels increased steadily (Figure 3.1). China focused on monitoring and inspection in its efforts to build capacity. During the period between 2006 and 2008, the two functions accounted for 85% of government personnel growth for environmental protection (Ministry of Environmental Protection, 2006–2009).

Because of the concern about their data accuracy and reliability, as discovered in my interviews, CEMSs were not the only data source to track the operation of SO_2 scrubbers. Other relevant data were collected, including operation and maintenance records, load factors of electricity generation, sulfur contents of coal, the consumption of limestone and other reagents, electricity consumption, the handling of products from SO_2 scrubbers, the opening and closure of bypass dampers and records of accidents and responses (NDRC and SEPA, 2007b; SEPA, 2007). SO_2 concentration in the inlet flue gas corresponds to the sulfur contents within a fairly predictable range. The load factors of electricity generation decide the flow rate of the flue gas and can check direct measurement with CEMSs. The factors together determine the sulfur load to an SO_2 scrubber system. For wet scrubbers using limestone as the reagent, the molar ratio between $CaCO_3$ and SO_2 is normally quite stable at approximately 1.02 to 1.05 (Ministry of Environmental Protection, 2010b). Then the sulfur load would decide the consumption of limestone and the production of gypsum. The managers of coal-fired power plants were asked to keep the receipts of limestone purchases, and cheating on receipts was considered financial fraud, with harsh penalties on those responsible. Electricity is another important input to operate SO_2 scrubbers. Because all data should be consistent with each other, it became more difficult to cheat.

The problem of collusion appeared under control. Data from CEMSs were sent to more than one agency, including environmental protection bureaus and electric grid corporations. Authorities at China's four government levels – central, provincial, prefectural and county – all inspected SO_2 scrubbers. The multiplicity of inspection authorities effectively diminished the opportunities of collusion. In addition, the pressure to achieve the 10% reduction goal of SO_2 emissions in the 11th Five-Year Plan reduced incentives to collude.

4.3 Building the screening system with big data

The preceding measures to strengthen the diagnosing system indeed worked, but for achieving an even deeper reduction of SO_2 emissions, China faces much more daunting problems in dealing with smaller polluting sources that are a few orders of magnitude greater in numbers. New opportunities are emerging with newly emerged environmental compliance monitoring technologies (Kitchin,

2014), which are evolving rapidly in terms of effectiveness in catching noncompliance and efficiency in utilizing compliance monitoring resources. For example, CEMSs played a central role in the U.S. Acid Rain Program as well as the European Union Emission Trading Scheme (The U.S. Congress, 1990; Stranlund and Chavez, 2000; European Commission, 2012). Remote-sensing technologies using satellites could provide large-scale spatial coverage of multiple pollutants (Streets et al., 2013). The measurement extends to areas beyond the current monitoring network, although the spatial resolution is coarse (Streets et al., 2013). Social media and the prevalent use of smartphones have greatly facilitated and strengthened the power of the civil society in monitoring environmental pollution and compliance (Stevens and Ochab, 2010; Kay et al., 2015). Various types of sensors, in addition to novel carriers such as unmanned aerial vehicles, have been more and more widely adopted to measure pollution levels (Snyder et al., 2013; Wang and Brauer, 2014).

China has been actively seeking opportunities in big data that can be applied for environmental protection. In 2015, State Council formally issued the Action Outline for Promoting Big Data Development to encourage the wide integration of big data in governance (State Council, 2015). In 2016, the then Ministry of Environmental Protection enacted the Comprehensive Plan on Ecological and Environmental Big Data Construction (Ministry of Environmental Protection, 2016a). It listed a comprehensive plan on how big data could be collected, integrated, developed and applied for environmental compliance monitoring, enforcement and management.

These new compliance monitoring technologies shed light on new solutions to the old challenges. First, in addressing the compliance monitoring resource constraint, these technologies could potentially provide a relatively low-cost means to monitor polluting sources. For example, although one satellite observing the Earth's CO_2 and air quality could cost a few hundred million U.S. dollars, such as the OCO-2 satellite for CO_2 monitoring by the National Aeronautical and Space Administration with a price tag of US$465 million, its wide spatial and regular coverage would substantially reduce the average and, especially, marginal costs for one observation (Wall, July 2, 2014; Osterman et al., 2018). Second, many of these technologies could circumvent various levels of local governments and polluting sources to provide top-down, external and objective data without subjective distortions. They are originated from entirely different external sources, not internal reporting. Satellite or remote-sensing data could be gathered in a centralized manner without the direct involvement of local governments or polluting sources themselves.

Nevertheless, these new technologies also have a critical weakness. Most of them generally have not reached the minimum accuracy requirements to legally or administratively punish polluters, while conventional technologies currently in application (although not all) could fulfill the requirements if intentional data manipulation is effectively deterred. Remote-sensing data have been successfully applied in China to examine the impacts of environmental policies on pollutant emissions from coal-fired power plants, but the accuracy has not been adequate to justify their direct application in legally determining the compliance status of individual polluting firms (Zhang et al., 2009; Li et al., 2010).

The trade-offs between conventional and new technologies indicate that the latter cannot completely replace the former at their current stage, but their clear advantages in costs (and objectiveness) are crucial considerations for China's ongoing reform on the conventional diagnosing system to deeply integrate big data and other technologies. Section 4.4 mainly focuses on how this reform may achieve better efficiency and effectiveness in environmental compliance monitoring. Different technologies are recognized to have different features mainly from cost and accuracy perspectives. Their weaknesses and strengths could complement each other for building a better system than any individual category of technologies can do alone.

4.4 Comparing diagnosing and screening systems

Environmental compliance rates are simulated with empirically defined parameters as discussed in the Appendix to this chapter. This subsection discusses the model simulation and sensitivity analysis results. If any input parameter is not targeted in a simulation, it will adopt the empirical value as specified in the current scenario as summarized in Table 6.3.

Compliance rates $(1 - M^t)$ in the screening system depend on their initial levels $(1 - M^0$; Figure 6.3). For example, if initially with 40,000 inspection staff, the screening system results in two equilibrium compliance rates $(1 - M^*)$ after several time steps, about 27% (a very low compliance rate) and 100% (full compliance; Figure 6.3). An equilibrium state is defined as, given the empirical values of parameters, the compliance rate remains stable over time and swings back if a small disturbance happens (Table 6.3). When the initial compliance rates are above a certain level, the final equilibrium compliance rates tend to converge to a high level close to full compliance. However, when the initial compliance rates are below that level, the available resources would not be adequate to catch enough noncompliance cases. Noncompliance will become the dominant choice of rational polluters, or the compliance monitoring system falls into a noncompliance trap due to its equilibrium status. The following simulations of the screening system will primarily report equilibrium compliance rates. In contrast, the diagnosing system demonstrates no memory. Its compliance rates at each time-step $(1 - M^t)$ have no relationship with the initial or proceeding levels $(1 - M^0$ and $1 - M^{t-1})$. They are decided only by immediately available compliance monitoring resources $(R^t$; Figure 6.3).

The relative effectiveness of the diagnosing and screening systems in enhancing compliance rates depends heavily on resource availability $(R^t$; Figure 6.4). When resources were too scarce (e.g., less than 30,000 inspection staff or half of China's available personnel in 2015), neither system would be able to result in high-compliance statuses, although the diagnosing system could achieve slightly better outcomes. When resources were abundant (more than 130,000 inspection staff or doubling the available personnel in 2015), either system would lead to

Table 6.3 Key parameters in the model and their empirical values

Parameters	Empirical values in the current scenario	Empirical ranges in the model simulation	Simulation
M^t	Noncompliance rate (%) at time-step t; the corresponding compliance rate is $1 - M^t$. Equilibrium noncompliance rate, $M^*_.$, is defined as the level when $M^t - M^{t-1} = 0.0\%$.	Initial noncompliance rate, $M^0_.$, is assumed to be 50% as the middle point of the full possible range between 0% and 100%.	M^0 varies from 0% (full compliance) to 100% (complete noncompliance), which covers the full range of possible compliance rates	Figure 6.3
R^t or R	Total available resources for compliance monitoring at the time-step t, which could be allocated between screening and diagnosing, R^t_s and R^t_d; R^t can be changed exogenously at a time step. When it remains a constant: $R^t = R$.	66,379 inspection staff (in 2015; Ministry of Environmental Protection, 2002–2016), within which 46,800, or 70.5%, were environmental inspectors (in 2017) as in the "double randomness, one publicization" databases (Ministry of Ecology and Environment, 2018)	From 1,959 inspection staff (the total number of inspection staff at the central and provincial levels in 2015) to 185,108 (the total number of environmental officials at all levels for administration, inspection and monitoring in 2015; Ministry of Environmental Protection, 2002–2016)	Figure 6.4
N	The number of polluting sources under compliance monitoring	809,500 polluters under compliance monitoring as in the "double randomness, one publicization" databases (Ministry of Ecology and Environment, 2018)	From 68,121 polluting sources that were under special supervisory monitoring in 2015 (Ministry of Environmental Protection, 2002–2016) to 5,925,576 in China's first census of polluting sources with the census date being December 31, 2007 (Ministry of Environmental Protection et al., 2010)	Figure 6.5(a)
$\dfrac{C}{P}$	C: pollution abatement costs (US$/ton), which vary across polluting sources; P: the penalty on noncompliance (US$/ton), which is assumed to be fixed for every punished polluting source.	2/3 (P is assumed to be 1.5 times of the pollution abatement costs, being a middle ground in China's empirical cases as introduced in the cell to the right)	From 0.1 (very harsh P; in the 2007 regulation for operating SO_2 scrubbers, P was five times of C [Xu, 2011a; NDRC and SEPA, 2007b]) to 1.6 (very lenient P with C 60% higher than P; in China's shale-gas development, the then P was significantly lower than C [Guo et al., 2014])	Figure 6.5(b)

$\Phi(\bullet)$	Cumulative distribution function of $\frac{C}{P}$	A normal distribution is assumed with a standard deviation of 0.33.	The impacts of a lognormal distribution are also simulated, due to the lack of actual information.	Figure 6.5(c)
r	Required resources to screen and diagnose *one* polluting source (r_s and r_d, respectively), which are assumed to remain unchanged over time	r_d : 0.074 inspector-year per inspection (see text for empirical estimation); $\frac{r_s}{r_d}$ is assumed to be 10%, or r_s is equivalently 0.0074 inspector-year per inspection.	Due to inadequate information, the ratio, $\frac{r_s}{r_d}$, is examined with a full possible range from 1% to 100%. By definition, screening technologies must be cheaper than diagnosing technologies. Otherwise, the latter will be better from both cost and accuracy perspectives to make the former obsolete.	Figure 6.5(d)
K_1	The probability (%) that one technology recognizes compliant cases as being compliant; $1 - K_1$: (Type I error) the probability that compliant cases are recognized as being noncompliant.	K_{1s} and K_{1d}, the corresponding probabilities for screening and diagnosing technologies, are assumed to be 90% and 99%, respectively.	Due to inadequate information, a full range, 0%~100%, is examined.	Figure 6.6(a) & (b)
K_2	The probability (%) that one technology recognizes noncompliant cases as being noncompliant; $1 - K_2$: (Type II error) the probability that noncompliant cases are recognized as being compliant	K_{2s} and K_{2d}, the corresponding probabilities for screening and diagnosing technologies, are assumed to be 70% and 90%, respectively.	Due to inadequate information, a full range, 0%~100%, is examined.	Figure 6.6(c) & (d)

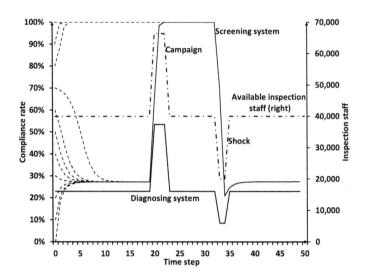

Figure 6.3 Model simulation of compliance rates $(1 - M^t)$ in the diagnosing and screening systems with available compliance monitoring resources (i.e., exogenously determined number of inspection staff in the dashed curve, R^t) and initial compliance rates $(1 - M^0$, from 0% to 100%)

Note: After compliance rates reach equilibrium levels $(1 - M^* 1 - M^0 1 - M^t)$, a hypothetical environmental campaign (temporarily with more inspection staff) is exogenously triggered to run for three time-steps and a hypothetical shock (temporarily with fewer inspection staff) for two time-steps. Their periods are indicated alongside the dashed curve. All other model parameters adopt the empirical values in the current scenario in Table 6.3.

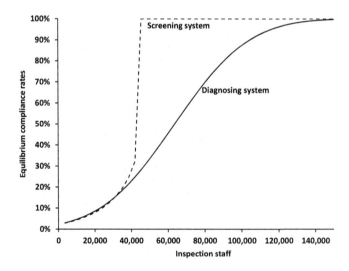

Figure 6.4 Model simulation of equilibrium compliance rates $(1 - M^*)$ in the screening and diagnosing systems in relation to available inspection staff (R)

nearly full compliance and strategies would not matter much. Most situations in the real world, including China in 2015 with 66,379 inspection staff, should fall in between: resources are constrained but neither unlimited nor depleted. In these situations, the two systems would diverge away from each other and the screening system could use available resources much more efficiently to achieve significantly higher compliance rates (Figure 6.4).

The number of polluters (N) matters greatly for the relative performance of the two compliance monitoring systems. With 66,379 environmental inspection staff in the current scenario, the screening system shows significantly higher compliance rates than the diagnosing system when the number of polluters is between 0.5 million to about 1.2 million (Figure 6.5(a)). Both systems could effectively handle fewer than 0.5 million polluters for their nearly full compliance, while neither system could be up for the job with more than 1.2 million polluters. As discussed in the Appendix at the end of this chapter, the polluting sources under the central government's special supervisory monitoring, generally large or hazardous polluters, were 68,121 in 2015. The currently available inspection staff would be of little resource constraint to achieve their general environmental compliance, as in China's current situation. The "double randomness, one publicization" scheme covered 809,500 polluters, for which the screening system with nearly full compliance tends to have a great advantage over the diagnosing system with only about half of polluters under compliance. If compliance monitoring does not differentiate the 5,925,576 polluting sources in the 2007 census, the overall compliance rate would be very low, being less than 5%. As in the Chinese practice, compliance monitoring should strategically allocate resources to those bigger and more severe polluters. Otherwise, the system would be overwhelmed. From another perspective, the model simulation also indicates that small polluters have significantly low environmental compliance rates.

As enlightened in the economic theory of crime and punishment, a penalty could enhance compliance rates in a similar way as compliance monitoring. When the penalty level is ten times of the pollution abatement costs (i.e., $\frac{C}{P}$ being 0.1), both the screening and the diagnosing systems could yield nearly full compliance (Figure 6.5(b)). For example, in ensuring the normal operation of SO_2 scrubbers, the penalty for noncompliance was five times the pollution abatement costs (i.e., $\frac{C}{P}$ being 0.2; Xu, 2011a). China's compliance monitoring system was closer to the diagnosing system, but it still effectively brought coal-fired power plants under prevalent compliance as projected by the model (Figure 6.5(b); Xu, 2011a). When the penalty level barely catches up with the abatement costs (i.e., $\frac{C}{P} > 1$), neither system would work although the screening system performs even worse (Figure 6.5(b)). This was the case in China's early days in dealing with water pollution in shale-gas development (Guo et al., 2014).

A deviation of the statistical distribution of the cost/penalty ratio ($\Phi(\bullet)$) does not seem to cause much difference for the earlier simulation results

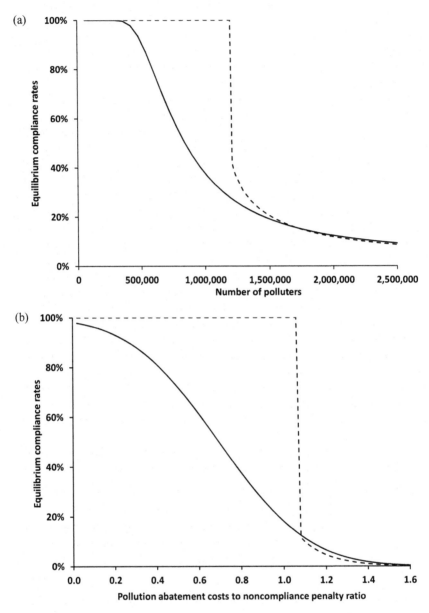

Figure 6.5 Model simulation of equilibrium compliance rates $(1-M^*)$ in the screening and diagnosing systems in relation to (a) the number of polluters (N); (b) the ratios between pollution abatement costs and noncompliance penalty $\left(\dfrac{C}{P}\right)$; (c) available inspection staff (R), where the pollution abatement cost-to-noncompliance penalty ratio $\left(\dfrac{C}{P}\right)$ has a *lognormal distribution* $(\Phi(\bullet))$; and (d) the relative resource intensity of screening and diagnosing technologies $\left(\dfrac{r_s}{r_d}\right)$

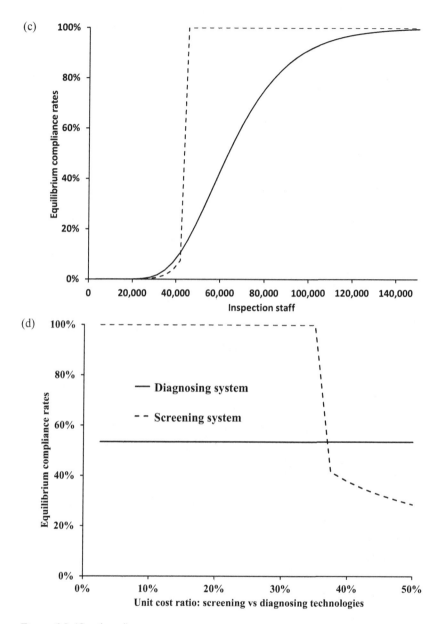

Figure 6.5 (Continued)

(Figure 6.5(c)). When the pollution abatement cost-to-noncompliance penalty ratio $\left(\dfrac{C}{P}\right)$ is assumed to have a lognormal distribution, the relationship between available inspection staff and equilibrium compliance rates is similar to the situation earlier (Figure 6.5(c)).

Screening technologies should be carefully selected. Otherwise, the screening system would not yield higher compliance rates than the diagnosing system. The low costs of a screening technology to monitor one polluting source (r_s) is crucial for its better performance (Figure 6.5(d)). It should be no less than 65% cheaper than a diagnosing technology (r_d; Figure 6.5(d)). Furthermore, in terms of accuracy, compliance monitoring technologies for screening and diagnosing have distinctly different requirements on their Type I and II errors. Screening technologies should make fewer Type I errors in wrongly recognizing compliant cases into the high-risk group (K_{1s} should be generally above 60%; Figure 6.6(a)), while diagnosing technologies should make fewer Type II errors in wrongly recognizing noncompliant cases as being compliant for them to evade penalties (K_{2d} must be generally above 60%; Figure 6.6(d)). The requirements on the other two accuracy indicators are much more relaxed. Screening technologies should not put more than 80% of noncompliant cases into the low-risk group (K_{2s} must be generally above 20%; Figure 6.6(c)). The probability of a diagnosing technology to recognize compliant cases as being compliant seems to matter little (K_{1d}; Figure 6.6(b)). Although this model assumes only two compliance statuses of a polluter, being compliant or noncompliant, polluters do differ in terms of the noncompliance severity. In the terminology of this model, compliance monitoring technologies should inherently have thresholds on whether to recognize a polluter as being compliant or not. The preceding accuracy indicators, especially K_{2s} and K_{2d}, also reflect such thresholds. Accordingly, screening technologies only need to catch those more severe noncompliant cases or with strong noncompliant signals (due to the relaxed requirement of K_{2s}) while diagnosing technologies must convict most of these severe noncompliant polluters (K_{2d}). These results could serve as the guideline for assessing and selecting screening and diagnosing technologies.

Overall, the screening system in general does show significantly better performance than the diagnosing system to achieve higher compliance rates. Depending on initial compliance rates and available resources, compliance rates in the screening system may evolve into two equilibrium levels, being at nearly full compliance and prevalent noncompliance. At the 2015 level of inspection staff in China, the screening system would be able to yield nearly full compliance for the 809,500 polluting sources as covered under the "double randomness, one publicization" scheme. However, the diagnosing system that is closer to reality would only bring about half of those polluters under compliance.

4.5 Resilience of screening and diagnosing systems

Campaigns or movements (*yundong*) are widely used in China's governance. In order to achieve a highly prioritized goal within a short time, the government may intensively reallocate unusual amounts of human, financial or political resources for certain tasks. These resources are usually "borrowed" from other agencies or functions and thus must be "returned" after campaigns conclude. Examples include anticrime campaigns, especially "strike hard" (Trevaskes, 2010); anticorruption campaigns (Wedeman, 2005); and environmental campaigns (Jahiel, 1998; van

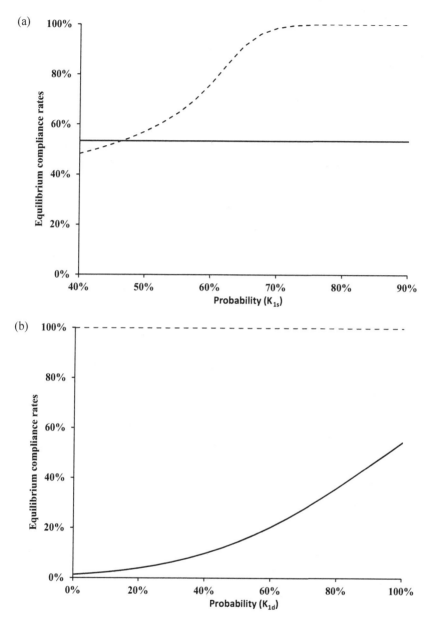

Figure 6.6 Model simulation of equilibrium compliance rates $(1 - M^{*})$ in the screening and diagnosing systems in relation to the probabilities that (a) the screening technology recognizes compliance cases as being compliant (K_{1s}), (b) the diagnosing technology recognizes compliance cases as being compliant (K_{1d}), (c) the screening technology recognizes noncompliance cases as being noncompliant (K_{2s}) and (d) the diagnosing technology recognizes noncompliance cases as being noncompliant (K_{2d})

(c)

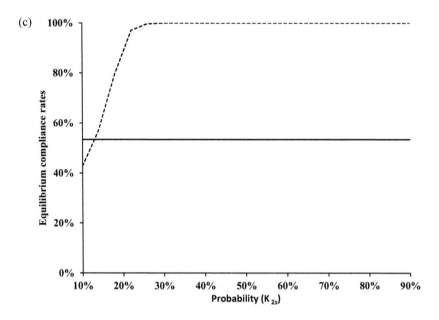

(d)

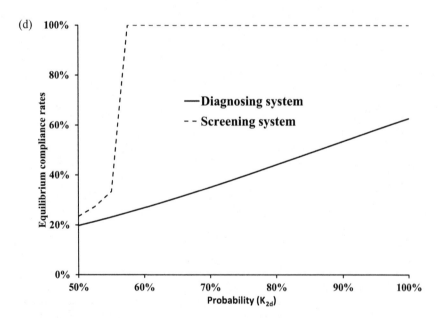

Figure 6.6 (Continued)

Rooij, 2006). Opposite to environmental campaigns, compliance monitoring might also experience shocks as observed in the author's fieldwork in China, for example, when inspection staff in one region or for one environmental task are temporarily "borrowed" for launching campaigns in another region or for other tasks.

Campaigns can achieve rapid progress on the targeted tasks. However, when the temporarily available resources are retreated, such campaign-style compliance monitoring and enforcement often fail to reach sustained compliance. One notable example of a largely short-lived environmental enforcement campaign is the "midnight action" for solving the unacceptable water pollution in the Huai River in 1997 that shut down about 5,000 small polluting factories (Bai and Shi, 2006; Liu, 1998). Improvements were achieved in the short term with significantly reduced water pollutant emissions and cleaner water quality (Liu, 1998). However, pollution rebounded quickly after the campaign was over (Bai and Shi, 2006).

The compliance monitoring model as constructed in this study provides an understanding of the short-lived impacts of environmental compliance monitoring campaigns and shocks in the diagnosing system. The diagnosing system has no memory, and its compliance rate at a given time directly corresponds to the immediately available enforcement resources (Figure 6.3). In contrast, the screening system has a memory and this feature suggests that short-term environmental campaigns might be more strategically utilized to establish the screening system for compliance monitoring and achieve high compliance rates. As illustrated in Figure 6.3, although compliance monitoring resources are kept at the same level before and after environmental campaigns, the equilibrium compliance rate will be fundamentally lifted from a low to a high status. The compliance rate evolution could be explained with the compliance monitoring model. Before a campaign starts, the prevalent noncompliance indicates that a great majority of compliance monitoring resources should be spent in the diagnosing step to convict polluters. A small proportion of resources will be enough to screen noncompliant cases for the relatively expensive diagnosing. When the environmental campaign is launched, with more and more noncompliant polluters being caught in noncompliance, their rational decisions will result in higher compliance rates. Then fewer polluting sources will be screened into the high-risk group in the following time-step, which requires less resource for diagnosing. In addition, the simulation also suggests that if transformed into the screening system, China might reduce the number of environmental inspection staff from the current level but still maintain high compliance rates.

Environmental compliance monitoring shocks have opposite impacts as campaigns. A temporary shortage of inspection staff could destabilize high equilibrium compliance rates back to low levels (Figure 6.3). As explained in the model construction, the compliance rate in the screening system at a time-step is only affected by that in the previous time-step. This short memory leads to the screening system's limited resilience when facing environmental compliance monitoring shocks.

If compliance rates with a longer past contribute to compliance decisions at a current time-step, the screening system of compliance monitoring will become more resilient. A longer memory shows that environmental campaigns should run longer for elevating the compliance rate to a higher equilibrium, while temporary environmental shocks would be less damaging, with the dipped compliance rate quickly rebounding afterward.

Environmental campaigns with temporary increases in inspection staff or shocks with their temporary reduction could destabilize the equilibrium rates in the screening system with longer-term impacts, while the impacts in the diagnosing system would be short-lived as seen in empirical cases. Environmental campaigns might be especially utilized to pull the system out of a possible non-compliance trap. If polluters have longer memories and their current compliance rate is directly determined by those in the past multiple periods, the screening system will demonstrate more resilience against the short-term campaigns and shocks.

Note

1 Adapted with permission from XU, Y. 2011. Improvements in the operation of SO_2 scrubbers in China's coal power plants. *Environmental Science & Technology*, 45, 380–385. Copyright (2011) American Chemical Society. Much has been revised and expanded on.

References

Arguedas, C. 2008. To comply or not to comply? Pollution standard setting under costly monitoring and sanctioning. *Environmental & Resource Economics*, 41, 155–168.

Bai, X. & Shi, P. 2006. Pollution control: In China's Huai river basin: What lessons for sustainability? *Environment: Science and Policy for Sustainable Development*, 48, 22–38.

Becker, G. S. 1968. Crime and punishment – economic approach. *Journal of Political Economy*, 76, 169–217.

Blackman, A. & Harrington, W. 2000. The use of economic incentives in developing countries: Lessons From international experience with industrial air pollution. *The Journal of Environment Development*, 9, 5–44.

Bonita, R., Beaglehole, R. & Kjellström, T. 2006. *Basic epidemiology*. Washington, DC: WHO.

Cao, D., Yang, J. & Ge, C. 1999. SO_2 charge and tax policies in China: Experiment and reform. *In: Environmental taxes: Recent developments in China and OECD countries.* Paris, France: OECD.

Chenery, H. B. 1961. Comparative advantage and development policy. *American Economic Review*, 51, 18–51.

Cohen, M. A. 1999. Monitoring and enforcement of environmental policy. *In:* Folmer, H. & Tietenberg, T. (eds.) *The international yearbook of environmental and resource economics 1999/2000.* Cheltenham: Edward Elgar Publishing Limited.

Dasgupta, S., Wang, H. & Wheeler, D. 1997. *Surviving success: Policy reform and the future of industrial pollution in China.* World Bank Policy Research Working Paper 1856. Washington, DC: World Bank Group.

The Economic & Trade Commission of Jiangsu Province. 2009. *Quarterly thermal efficiencies of electricity generation units.* Nanjing, China.

EPA. 2003. *Air pollution control technology fact sheet – flue gas desulfurization.* EPA-452/F-03-034. Washington, DC: EPA.

EPA & DOE. 2010. *Electric power annual 2008.* Washington, DC: EPA.

European Commission. 2012. *The monitoring and reporting regulation – general guidance for installations* [Online]. Available: https://ec.europa.eu/clima/sites/clima/files/ets/monitoring/docs/gd1_guidance_installations_en.pdf.

Glaeser, E. L. 1999. *An overview of crime and punishment.* Cambridge, MA: Harvard University and NBER.

Gray, W. B. & Deily, M. E. 1996. Compliance and enforcement: Air pollution regulation in the US steel industry. *Journal of Environmental Economics and Management,* 31, 96–111.

Guo, M. Y., Xu, Y. & Chen, Y. Q. D. 2014. Fracking and pollution: Can China rescue its environment in time? *Environmental Science & Technology,* 48, 891–892.

Heindl, P. 2012. *Transaction costs and tradable permits: Empirical evidence from the EU emissions trading scheme.* ZEW Discussion Paper No. 12-021 [Online]. Available: https://www.econstor.eu/handle/10419/56029.

Helland, E. 1998. The enforcement of pollution control laws: Inspections, violations, and self-reporting. *Review of Economics and Statistics,* 80, 141–153.

Henan Department of Environmental Protection. 2010. *Information on the collection of SO_2 effluent discharge fee from province-regulated coal-fired power plants in the first quarter of 2010.* Zhenzhou, China: Henan Department of Environmental Protection.

Henan Development and Reform Commission & Henan Environmental Protection Bureau. 2007. *A notice to transfer NDRC and SEPA's policy: Management measures on desulfurized electricity price and the operation of desulfurization facilities of coal-fired power generators (on trial).* Zhenzhou, China: NDRC and SEPA.

IEA. 2009. *Cleaner coal in China.* Paris, France: IEA.

Jahiel, A. R. 1998. The organization of environmental protection in China. *The China Quarterly,* 156, 757–787.

Jiangsu Department of Environmental Protection. 2007–2009. *Monthly report on the operation of SO_2 scrubbers at coal-fired power plants.* Nanjing, China: Jiangsu Department of Environmental Protection.

Jiangsu Department of Environmental Protection. 2008. *On strengthening the collection of effluent discharge fee from coal-fired power plants.* Nanjing, China: Jiangsu Department of Environmental Protection.

Jin, Y. N., Andersson, H. & Zhang, S. Q. 2016. Air pollution control policies in China: A retrospective and prospects. *International Journal of Environmental Research and Public Health,* 13.

Kay, S., Zhao, B. & Sui, D. 2015. Can social media clear the air? A case study of the air pollution problem in Chinese cities. *Professional Geographer,* 67, 351–363.

Kitchin, R. 2014. *The data revolution.* Thousand Oaks, CA: Sage Publications.

Konrad, K. A., Lohse, T. & Qari, S. 2017. Compliance with endogenous audit probabilities. *Scandinavian Journal of Economics,* 119, 821–850.

Kruger, J. & Egenhofer, C. 2006. Confidence through compliance in emissions trading markets. *Sustainable Development Law & Policy,* 6, 2–13, 63–64.

Levitt, S. D. 2004. Understanding why crime fell in the 1990s: Four factors that explain the decline and six that do not. *Journal of Economic Perspectives,* 18, 163–190.

Li, C., Zhang, Q., Krotkov, N. A., Streets, D. G., He, K. B., Tsay, S. C. & Gleason, J. F. 2010. Recent large reduction in sulfur dioxide emissions from Chinese power plants observed by the ozone monitoring instrument. *Geophysical Research Letters,* 37.

Lin, J. Y., Cai, F. & Li, Z. 2003. *The China miracle: Development strategy and economic reform.* Sha Tin: Published for the Hong Kong Centre for Economic Research and the International Center for Economic Growth by the Chinese University Press.

Liu, H. 1998. An overview of water pollution prevention and control in the Huai River valley. *Enviornmental Management in China*, 5–8.

Liu, Z., Guan, D. B., Wei, W., Davis, S. J., Ciais, P., Bai, J., Peng, S. S., Zhang, Q., Hubacek, K., Marland, G., Andres, R. J., Crawford-Brown, D., Lin, J. T., Zhao, H. Y., Hong, C. P., Boden, T. A., Feng, K. S., Peters, G. P., Xi, F. M., Liu, J. G., Li, Y., Zhao, Y., Zeng, N. & He, K. B. 2015. Reduced carbon emission estimates from fossil fuel combustion and cement production in China. *Nature*, 524, 335.

Lu, X., Dudek, D. J., Qin, H., Zhang, J., Lin, H., Yang, Z. & Wang, Y. 2006. Survey on the capacity of environmental administrative enforcement in China. *Research of Environmental Sciences*, 19, 1–11.

Lu, Z., Streets, D. G., Zhang, Q., Wang, S., Carmichael, G. R., Cheng, Y. F., Wei, C., Chin, M., Diehl, T. & Tan, Q. 2010. Sulfur dioxide emissions in China and sulfur trends in East Asia since 2000. *Atmospheric Chemistry and Physics*, 10, 6311–6331.

Lu, Z., Zhang, Q. & Streets, D. G. 2011. Sulfur dioxide and primary carbonaceous aerosol emissions in China and India, 1996–2010. *Atmospheric Chemistry and Physics*, 11, 9839–9864.

McAllister, L. K., Van Rooij, B. & Kagan, R. A. 2010. Reorienting regulation: Pollution enforcement in industrializing countries. *Law & Policy*, 32, 1–13.

MEP. 2013. *The 12th five-year plan on capacity building of environmental regulations.* Beijing, China: MEP.

MEP & AQSIQ. 2011. *Emission standard of air pollutants for thermal power plants.* Beijing, China: MEP, AQSIQ.

MEP & NDRC. 2008. *Announcement on punishing coal-fired power plants with abnormal operation of SO_2 scrubbers in 2007.* Beijing, China: MEP, NDRC.

Ministry of Ecology and Environment. 2018. *The comprehensive implementation of 'double randomnesses, one publicization' for environmental inspection.* Beijing, China: Ministry of Ecology and Environment.

Ministry of Environmental Protection. 2002–2016. *Annual statistical report on the environment in China.* Beijing, China: Ministry of Environmental Protection.

Ministry of Environmental Protection. 2006–2009. *Annual statistical report on the environment in China.* Beijing, China: Ministry of Environmental Protection.

Ministry of Environmental Protection. 2008. *Statement to penalize coal-fired power plants for the abnormal operation of their SO_2 scrubbers in 2007.* Beijing, China: Ministry of Environmental Protection.

Ministry of Environmental Protection. 2009a. *2008 assessment report on provincial major pollutant emissions.* Beijing, China: Ministry of Environmental Protection.

Ministry of Environmental Protection. 2009b. *Information on the mitigation of major pollutants in 2008.* Beijing, China: Ministry of Environmental Protection.

Ministry of Environmental Protection. 2009c. *Statement to penalize five coal-fired power plants for the abnormal operation of their SO_2 scrubbers in 2008.* Beijing, China: Ministry of Environmental Protection.

Ministry of Environmental Protection. 2010a. *China's capacities of water treatment plants and SO_2 scrubbers at coal-fired power plants.* Beijing, China: Ministry of Environmental Protection.

Ministry of Environmental Protection. 2010b. *Guideline on best available technologies of pollution prevention and control for coal-fired power plant industry (on trial).* Beijing, China: Ministry of Environmental Protection.

Ministry of Environmental Protection. 2016a. *Comprehensive plan on ecological and environmental big data construction.* Beijing, China: Ministry of Environmental Protection.

Ministry of Environmental Protection. 2016b. *final accounts of ministry of environmental protection*. Beijing, China: Ministry of Environmental Protection.

Ministry of Environmental Protection, National Statistics Bureau & Ministry of Agriculture. 2010. *Public report on the first national census of polluting sources*. Beijing, China: Ministry of Environmental Protection.

National Bureau of Statistics. 2008. *China energy statistical yearbook*. Beijing, China: China Statistics Press.

NDRC. 2004. *Technical code for designing flue gas desulfurization plants of fossil fuel power plants*. DL/T5196-2004. Beijing, China: NDRC.

NDRC & SEPA. 2007a. *The 11th five-year plan on SO2 control in existing coal-fired power plants*. Beijing, China: State Environmental Protection Administration, NDRC.

NDRC & SEPA. 2007b. *Management measures on desulfurized electricity price permium and the operation of desulfurization facilities in coal-fired power generators (on trial)*. Beijing, China: State Environmental Protection Administration, NDRC.

OECD. 2006. *Environmental compliance and enforcement in China – an assessment of current practices and ways forward*. Paris, France: OECD.

Ohlin, B. 1967. *Interregional and international trade*. Cambridge, MA: Harvard University Press.

Osterman, G., Eldering, A., Avis, C., Chafin, B., O'Dell, C., Frankenberg, C., Fisher, B., Mandrake, L., Wunch, D., Granat, R. & Crisp, D. 2018. *Orbiting carbon observatory – 2 (OCO-2) data product user's guide, operational L1 and L2 data versions 8 and lite file version 9*. Pasadena, CA: Jet Propulsion Laboratory, California Institute of Technology.

Pan, L., Wang, Z. & Wang, Z. 2005. Present status and countermeasure suggestion for thermal power plants CEMS in China. *Research of Environmental Sciences*, 18, 42–45.

Polinsky, A. M. & Shavell, S. 2000. The economic theory of public enforcement of law. *Journal of Economic Literature*, 38, 45–76.

Raufer, R. & Li, S. Y. 2009. Emissions trading in China: A conceptual 'leapfrog' approach? *Energy*, 34, 904–912.

Ricardo, D. 1817. *On the principles of political economy and taxation*. London: J. Murray.

Russell, C. S. & Vaughan, W. J. 2003. The choice of pollution control policy instruments in developing countries: Arguments, evidence and suggestions. *In:* Folmer, H. & Tietenberg, T. (eds.) *International yearbook of environmental and resource economics*. Cheltenham: Edward Elgar.

SEPA. 2006. *Guidelines on calculating SO₂ emission quotas*. Beijing, China: State Environmental Protection Administration.

SEPA. 2007. *Verification of major pollutants emission reduction in the 11th five-year period (on trial)*. Beijing, China: State Environmental Protection Administration.

SERC. 2009. *New credit to CEMSs at SO₂ scrubbers in Jiangsu province*. Nanjing, China: SERC.

SERC, NDRC, National Energy Bureau & MEP. 2009. *Report on electric industry's energy conservation and pollutant mitigation in 2008*. Beijing, China: SERC, NDRC, MEP.

Shimshack, J. P. 2014. The economics of environmental monitoring and enforcement. *Annual Review of Resource Economics*, 6, 339–360.

Snyder, E. G., Watkins, T. H., Solomon, P. A., Thoma, E. D., Williams, R. W., Hagler, G. S. W., Shelow, D., Hindin, D. A., Kilaru, V. J. & Preuss, P. W. 2013. The changing paradigm of air pollution monitoring. *Environmental Science & Technology*, 47, 11369–11377.

State Commission Office for Public Sector Reform. 2018. *Regulations on the authorities, organization and personnel of the ministry of ecology and environment*. Beijing, China: State Commission Office.

State Council. 2007a. *Notice on distributing composite working plan on energy conservation and pollutant emission reduction.* Beijing, China: State Council.

State Council. 2007b. *Notice on distributing implementation plans and methods of statistics, monitoring and assessment on energy conservation and pollutant emission reduction.* Beijing, China: State Council.

State Council. 2015. *Action outline for promoting big data development.* Beijing, China: State Council.

State Council. 2019. Advice on the comprehensive implementation of joint 'double randomnesses, one publicization'. *In: Market regulation.* Beijing, China: State Council.

State Development Planning Commission, Ministry of Finance, SEPA & State Economic and Trade Commission. 2003. *Measures for the administration of the charging rates for pollutant discharge fees.* Decree No. 31 [Online]. Available: http://www.lawinfochina. com/display.aspx?id=2705&lib=law&EncodingName=big5.

State Electricity Regulation Commission (Nanjing Office). 2009. *Information on the operation of SO2 scrubbers in coal-fired power plants in Jiangsu Province.* Nanjing, China: SERC.

State Environmental Protection Administration & General Administration of Quality Supervision Inspection and Quarantine. 2003. *Emission standard of air pollutants for thermal power plants.* GB 13223-2003. Beijing, China: State Environmental Protection Administration.

Steinfeld, E. S., Lester, R. K. & Cunningham, E. A. 2009. Greener plants, grayer skies? A report from the front lines of China's energy sector. *Energy Policy*, 37, 1809–1824.

Stevens, M. & Ochab, B. 2010. Participatory noise pollution monitoring using mobile phones. *Information Polity: The International Journal of Government & Democracy in the Information Age*, 15, 51–71.

Stranlund, J. K. & Chavez, C. A. 2000. Effective enforcement of a transferable emissions permit system with a self-reporting requirement. *Journal of Regulatory Economics*, 18, 113–131.

Streets, D. G., Canty, T., Carmichael, G. R., De Foy, B., Dickerson, R. R., Duncan, B. N., Edwards, D. P., Haynes, J. A., Henze, D. K., Houyoux, M. R., Jacobi, D. J., Krotkov, N. A., Lamsal, L. N., Liu, Y., Lu, Z. F., Martini, R. V., Pfister, G. G., Pinder, R. W., Salawitch, R. J. & Wechti, K. J. 2013. Emissions estimation from satellite retrievals: A review of current capability. *Atmospheric Environment*, 77, 1011–1042.

Trevaskes, S. 2010. *Policing serious crime in China: From 'strike hard' to 'kill fewer'.* London and New York: Routledge.

Tsinghua University. 2010. *A study on the management system of environmental pollution data collection in China.* Beijing, China: Tsinghua University Press.

The U.S. Congress. 1990. *Clean air act amendments 1990.* Washington, DC: The U.S. Congress.

The U.S. EPA. 2007. *CEMS cost model.* Washington, DC [Online]. Available: www.epa. gov/ttn/emc/cem/cems.xls.

Van Rooij, B. 2006. Implementation of Chinese environmental law: Regular enforcement and political campaigns. *Development and Change*, 37, 57–74.

Vossler, C. A. & Gilpatric, S. M. 2018. Endogenous audits, uncertainty, and taxpayer assistance services: Theory and experiments. *Journal of Public Economics*, 165, 217–229.

Wall, M. 2014. *NASA launches satellite to monitor carbon dioxide,* July 2 [Online]. Available: www.space.com/26403-nasa-oco2-carbon-dioxide-satellite-launch.html.

Wang, A. & Brauer, M. 2014. *Review of next generation air monitors for air pollution* [Online]. Available: https://www.semanticscholar.org/paper/Review-of-Next-Generation-Air-Monitors-for-Air-Wang-Brauer/ca660dbe9fa087b2bea4080064c251914155c177.

Wedeman, A. 2005. Anticorruption campaigns and the intensification of corruption in China. *Journal of Contemporary China*, 14, 93–116.

Wu, J. & Qian, Y. 2007. The construction of FGD at Beilun power plant on its 5×600 MW$_e$ units. *Electrical Equipment*, 8, 105–107.

Xu, Y. 2011a. Improvements in the operation of SO2 scrubbers in China's coal-fired power plants. *Environmental Science & Technology*, 45, 380–385.

Xu, Y. 2011b. The use of a goal for SO2 mitigation planning and management in China's 11th five-year plan. *Journal of Environmental Planning and Management*, 54, 769–783.

Zhang, Q., Streets, D. G., Carmichael, G. R., He, K. B., Huo, H., Kannari, A., Klimont, Z., Park, I. S., Reddy, S., Fu, J. S., Chen, D., Duan, L., Lei, Y., Wang, L. T. & Yao, Z. L. 2009. Asian emissions in 2006 for the NASA INTEX-B mission. *Atmospheric Chemistry and Physics*, 9, 5131–5153.

Zhang, Q., Streets, D. G., He, K., Wang, Y., Richter, A., Burrows, J. P., Uno, I., Jang, C. J., Chen, D., Yao, Z. & Lei, Y. 2007. NO$_x$ emission trends for China, 1995–2004: The view from the ground and the view from space. *Journal of Geophysical Research-Atmospheres*, 112.

Zhejiang Bureau of Quality and Technical Supervision. 2007. *The quota & calculation method of coal consumption for generating station*. Hangzhou, China: Zhejiang Bureau of Quality and Technical Supervision.

Appendix

Modeling environmental compliance monitoring systems

Key parameters

This model's primary output is the compliance rate of polluters under compliance monitoring: M^t is the noncompliance rate at the end of the time-step t, while $1-M^t$ is the corresponding compliance rate. For simulating the evolution of compliance rates over time, the model is designed to follow time-steps. In each time-step, enforcement activities are first conducted to comprise compliance monitoring and penalty on noncompliance. The compliance rate at the end of the previous time-step could affect the subsequent performance of environmental compliance monitoring, that is, the probability of catching noncompliance. This probability is assumed to be commonly available information for all polluters. Based on the expected penalty and compliance costs, polluters make compliance decisions to yield an overall compliance rate at the end of the current time-step.

As listed in Table 6.3, the model has a series of input parameters, whose values are given exogenously. They fall into several major categories: (1) environmental compliance monitoring system, including initial noncompliance rate (M^0), the total available resources for compliance monitoring (R^t) and the number of polluting sources (N); (2) the ratio between pollution abatement costs and penalty on noncompliance $\left(\dfrac{C}{P}\right)$ as well as its distribution ($\Phi(\bullet)$); (3) compliance monitoring technologies, including required resources for monitoring one polluting source (r), the probability that one technology recognizes compliant cases as being compliant (K_1) and the probability that one technology recognizes noncompliant cases as being noncompliant (K_2). Screening and diagnosing technologies are further distinguished with subscripts s and d, respectively. The four parameters (K_{1s}, K_{1d}, K_{2s} and K_{2d}) are assumed to be specific for a given compliance monitoring technology and do not change over time and cases. Compliance-monitoring technologies and systems could make two types of errors in identifying noncompliance (Polinsky and Shavell, 2000; Bonita et al., 2006). We assume H$_0$: a polluter is under environmental compliance. A Type I error indicates that a polluting firm is under compliance, but the environmental compliance monitoring wrongly identifies the case as noncompliance to mistakenly punish it. A Type II error refers to the situation that although a polluter is not complying, the system wrongly recognizes it as being compliant. Accordingly, the illegal polluter walks away without penalty.

Both errors consequently lower the deterrence effect, which might lead to lower compliance rates.

The diagnosing system

In this diagnosing-only system, the probability of noncompliant polluting sources that are rightfully punished is: $\frac{R_d^t}{r_d} \times \frac{1}{N} \times K_{2d}$, while the probability of compliant polluting sources that are mistakenly punished is $\frac{R_d^t}{r_d} \times \frac{1}{N} \times (1 - K_{1d})$. A polluter will choose compliance when $C + P \times \frac{R_d^t}{r_d} \times \frac{1}{N} \times (1 - K_{1d}) < P \times \frac{R_d^t}{r_d} \times \frac{1}{N} \times K_{2d}$, or $\frac{C}{P} < \frac{R_d^t}{r_d} \times \frac{1}{N} \times (K_{2d} + K_{1d} - 1)$. Because all resources are devoted to diagnosing, $R^t = R_d^t$. One polluting source could be diagnosed more than once to potentially incur a penalty every time that it is caught noncompliance. Corresponding to available enforcement resources, the noncompliance rate will be

$$M^t = 1 - \Phi(\frac{R^t}{r_d} \times \frac{1}{N} \times (K_{2d} + K_{1d} - 1)) \qquad \text{Equation 6.1}$$

Because M^t is not related to M^{t-1}, the noncompliance rate under the diagnosing system will not show dynamic evolution over time when other factors remain unchanged.

The screening system

When resources are inadequate, some polluting sources may be neither screened nor diagnosed, while the optimal allocation of resources will make sure that all screened-out polluting sources in the high-risk group are diagnosed and no available resource is wasted. Due to the existence of Type I and II errors, each group contains compliant and noncompliant sources. The high-risk group in the time-step t will comprise $\frac{R_s^t}{r_s} \times M^{t-1} \times K_{2s}$ noncompliant polluting sources and $\frac{R_s^t}{r_s} \times (1 - M^{t-1}) \times (1 - K_{1s})$ compliant polluting sources. Accordingly, the noncompliance rate in the high-risk group is $M_h^t = \frac{M^{t-1} \times K_{2s}}{M^{t-1} \times K_{2s} + (1 - M^{t-1}) \times (1 - K_{1s})}$.

The low-risk group will contain all remaining polluting sources, including those screened out and those not screened, $N - \frac{R_s^t}{r_s} \times M^{t-1} \times K_{2s} - \frac{R_s^t}{r_s} \times (1 - M^{t-1}) \times (1 - K_{1s})$, or $(N - \frac{R_s^t}{r_s}) + \frac{R_s^t}{r_s} \times M^{t-1} \times (1 - K_{2s}) + \frac{R_s^t}{r_s} \times (1 - M^{t-1}) \times K_{1s}$. The number of noncompliant polluting sources is $N \times M^{t-1} - \frac{R_s^t}{r_s} \times M^{t-1} \times K_{2s}$. Then the noncompliance rate in the low-risk group is $M_l^t = \dfrac{N \times M^{t-1} - \frac{R_s^t}{r_s} \times M^{t-1} \times K_{2s}}{N - \frac{R_s^t}{r_s} \times M^{t-1} \times K_{2s} - \frac{R_s^t}{r_s} \times (1 - M^{t-1}) \times (1 - K_{1s})}$.

After diagnosing, the number of noncompliant polluting sources that are rightfully punished is $\dfrac{R_d^t}{r_d} \times M_h^t \times K_{2d}$. The probability of noncompliant pollut-ing sources that are rightfully punished is $\dfrac{\dfrac{R_d^t}{r_d} \times M_h^t \times K_{2d}}{N \times M^{t-1}} = \dfrac{R_d^t}{r_d} \times \dfrac{1}{N} \times \dfrac{M_h^t}{M^{t-1}} \times K_{2d}$. The number of compliant polluting sources that are mistakenly pun-ished is $\dfrac{R_d^t}{r_d} \times (1 - M_h^t) \times (1 - K_{1d})$, and the corresponding probability is $\dfrac{\dfrac{R_d^t}{r_d} \times (1 - M_h^t) \times (1 - K_{1d})}{N \times (1 - M^{t-1})} = \dfrac{R_d^t}{r_d} \times \dfrac{1}{N} \times \dfrac{1 - M_h^t}{1 - M^{t-1}} \times (1 - K_{1d})$. We assume that the two probabilities are known to all polluting sources for their following compliance decisions.

Thus, the expected compliance cost is $C + P \times \dfrac{R_d^t}{r_d} \times \dfrac{1}{N} \times \dfrac{1 - M_h^t}{1 - M^{t-1}} \times (1 - K_{1d})$, while the expected penalty on noncompliance is $P \times \dfrac{R_d^t}{r_d} \times \dfrac{1}{N} \times \dfrac{M_h^t}{M^{t-1}} \times K_{2d}$. For a decision of compliance, the former should be lower than the latter:

$$C + P \times \frac{R_d^t}{r_d} \times \frac{1}{N} \times \frac{1 - M_h^t}{1 - M^{t-1}} \times (1 - K_{1d}) < P \times \frac{R_d^t}{r_d} \times \frac{1}{N} \times \frac{M_h^t}{M^{t-1}} \times K_{2d}, \text{or} \ \frac{C}{P} < \frac{R_d^t}{r_d} \times \frac{1}{N} \times$$

$$\frac{K_{2s} \times K_{2d} - (1 - K_{1s}) \times (1 - K_{1d})}{M^{t-1} \times K_{2s} + (1 - M^{t-1}) \times (1 - K_{1s})}.$$

$R_{\min}^t$ is further defined as a threshold when all polluting sources have just been screened ($R_s^t = N \times r_s$), all polluting sources in the high-risk group are diagnosed ($R_d^t = (\dfrac{R_s^t}{r_s} \times M^{t-1} \times K_{2s} + \dfrac{R_s^t}{r_s} \times (1 - M^{t-1}) \times (1 - K_{1s})) \times r_d$) and all resources are uti-lized $R^t = R_s^t + R_d^t$. Then $R_{\min}^t = N \times (r_s + (M^{t-1} \times K_{2s} + (1 - M^{t-1}) \times (1 - K_{1s})) \times r_d)$.

When $R^t \leq R_{\min}^t$, $R_d^t = \dfrac{R^t}{(\dfrac{r_d}{r_s} \times (M^{t-1} \times K_{2s} + (1 - M^{t-1}) \times (1 - K_{1s})) + 1)}$. The com-

pliance condition is $\dfrac{C}{P} < \dfrac{R^t}{N} \times \dfrac{K_{2s} \times K_{2d} - (1 - K_{1s}) \times (1 - K_{1d})}{(r_s + (M^{t-1} \times K_{2s} + (1 - M^{t-1}) \times (1 - K_{1s})) \times r_d)}$.

Additional compliance monitoring resources beyond $R_{\min}^t$ will be devoted to diagnosing those polluting sources in the high-risk group. These sources could be diagnosed and punished once or multiple times. In this situation, $R_d^t = R^t - R_s^t = R^t - N \times r_s$.

Then corresponding to available enforcement resources, the noncompliance rate at the end of time-step t will be

$$\text{If } R^t \leq R_{\min}^t, \ M^t = 1 - \Phi \left(\frac{R^t}{N} \times \frac{K_{2s} \times K_{2d} - (1 - K_{1s}) \times (1 - K_{1d})}{(r_s + (M^{t-1} \times K_{2s} + (1 - M^{t-1}) \times (1 - K_{1s})) \times r_d)} \right); \text{Equation 6.2}$$

If $R^t > R^t_{min}$, $M^t = 1 - \Phi\left(\dfrac{R^t - N \times r_s}{N \times r_d} \times \dfrac{K_{2s} \times K_{2d} - (1 - K_{1s}) \times (1 - K_{1d})}{M^{t-1} \times K_{2s} + (1 - M^{t-1}) \times (1 - K_{1s})}\right)$. Equation 6.3

R^t_{min} will be greater if the noncompliance rate at the end of time-step *t-1*, M^{t-1}, is higher or screening and diagnosing are more resource-intensive with greater r_s and r_d. Given a certain amount of total emissions under regulation, smaller individual polluting sources will result in a greater number of polluting sources, *N*, for compliance monitoring and thus higher demand for resources. Because the noncompliance rate, M^t, changes over time, R^t_{min} will change accordingly.

More accurate compliance monitoring technologies ($K_{1s}, K_{1d}, K_{2s}, K_{2d} \rightarrow 1$) with lower costs for an average polluting source ($r_s, r_d \rightarrow 0$) tend to induce higher compliance rates. Various factors could affect the availability of enforcement resources per polluting source ($\dfrac{R^t}{N}$). The economy of scale in compliance monitoring could have two folds. On one hand, larger polluting sources could lead to an internal economy of scale because the required enforcement resources are more related to the number of sources. More enforcement resources, larger polluting sources and a smaller amount of total emissions will increase the resource availability indicator. Even if with the screening step or effective compliance monitoring strategy, the probability of catching enough noncompliance cannot be enhanced to a high enough level without sufficient enforcement resources. On the other hand, the geographical proximity of polluting sources could provide an external economy of scale. The sources could then be equivalently bundled and reduce the compliance monitoring costs for one polluting source.

Corresponding to their required features, screening technologies are less accurate but also less expensive than diagnosing technologies. They must have such trade-offs to fit in the expected complementary roles. If one technology were both cheaper and more accurate than the other, the latter technology would be entirely replaced by the former.

Input parameters in China's empirical case

In order to empirically illustrate and analyze the model, the input parameters will adopt empirical values from the Chinese context. A current scenario and the range of parameters are defined with the best available empirical data in China's current situation. They are briefly summarized in Table 6.3, and this subsection provides a more detailed explanation.

Available resources for compliance monitoring (R^t) are a key input parameter that this model focuses on. For simplicity, compliance-monitoring resources (R^t) and costs of screening and diagnosing technologies (r_s and r_d) are counted as the number of environmental inspection staff. China has been gradually increasing governmental employees for environmental inspection. The resource availability still faces constraints, but it does not fall into the situation of extreme scarcity. From 2001 to 2015, staff for environmental inspection grew from 37,934 to 66,379 (Ministry of Environmental Protection, 2002–2016). More important, with the full establishment of regional supervisory centers/bureaus in 2008 by the then Ministry of Environmental Protection, the central government has significantly strengthened its capacity of environmental inspection, accounting for 0.48% (294 employees) of inspection staff at all four levels in 2009 and 0.82% (542 employees) in 2015, up from 0.07% in 2008 (41 employees; Ministry of Environmental Protection, 2002–2016). Six regional Supervision Bureaus were allowed to have, in total, 240 formal employees for taking charge of supervision tasks within their jurisdictions (State Commission Office for Public Sector Reform, 2018). Not all staff employed in the inspection section are environmental inspectors, for example, to play supporting roles such as office work. In 2017, China had 46,800 environmental inspectors in the databases for "double randomness, one publicization" (Ministry of Ecology and Environment, 2018). The closest year with available data on inspection staff was 2015. Accordingly, about 70.5% of inspection staff were environmental inspectors. The empirical model simulation adopts this ratio to examine the impacts of resource availability on environmental compliance rates. The current scenario thus has 66,379 inspection staff, or 46,800 environmental inspectors. If not specified, they will remain unchanged over time.

The number of polluting sources (N) was been briefly described in Section 4.2. The current scenario takes the intermediate number, 809,500 polluting sources as targeted in 2017 under the "double randomness, one publicization" scheme.

Pollution abatement costs and the associated penalty for noncompliance range across sectors, technologies and severity of noncompliance. The ratio between compliance costs and penalty $\left(\dfrac{C}{P}\right)$ is a key variable in this compliance monitoring model. In 2007, in order to tackle the long-term problem of weak environmental policy enforcement, China not only subsidized those coal-fired power plants to normally operate their SO_2 scrubbers but, more important, also issued a penalty, being five times of the subsidy/costs on a per-kilowatt-hour basis (Xu, 2011a; NDRC and SEPA, 2007b). In dealing with potential noncompliance on water pollution and withdrawal, however, China's penalty was barely able to catch up with the pollution abatement costs (Guo et al., 2014). In the current scenario, the cost/penalty ratio is assumed to be 2/3. Furthermore, pollution abatement costs are not identical across polluting firms due to, for example, economy of scale, the sulfur content of coal and whether the pollution removal facility is a retrofit or built together with the main equipment. In compiling China's SO_2 emission inventory, Lu et al. (2011) assumed that the sulfur content had a normal distribution. The current scenario follows, due to the key influence of sulfur contents on SO_2 abatement costs, to assume that the cost/penalty ratio $\left(\dfrac{C}{P}\right)$ has a normal distribution ($\Phi(\bullet)$) among the polluting sources.

The costs of screening and diagnosing technologies are accounted as the required number of inspectors in a year per environmental observation, either screening or diagnosing inspection (inspector-year per observation, being noted as r_s and r_d, respectively). China has comprehensively established the "double randomness, one publicization" method for governmental, including environmental and other, inspections on firms (State Council, 2019). For environmental inspections, the method had been well established in 2017 (Ministry of Ecology and Environment, 2018). Under this method, polluting firms and environmental inspectors will both be randomly selected from databases, while the information will be publicized to the public. In 2017, 809,500 polluting firms and 46,800 environmental inspectors were included in the databases, while 632,600 environmental inspections were conducted (Ministry of Ecology and Environment, 2018). Accordingly, 27 inspections were conducted by an average inspector in 2017. According to the author's earlier fieldwork in China (Guo et al., 2014; Xu, 2011a), one inspection generally involves two inspectors. Thus, the cost of environmental inspection or diagnosing technology (r_d) was 0.074 inspector-year per inspection. It is adopted in the current scenario.

Different screening and diagnosing technologies have different cost structures. For example, a sophisticated satellite-based technology has very high initial capital costs, but its marginal costs of monitoring one more pixel are negligible. For example, OCO-2 cost US$465 million to set up, but with more than 100,000 measurements of column CO_2 concentrations each day (Osterman et al., 2018), each measurement since its launch in July 2014 cost merely about US$2 to US$3 per measurement, considering neither operation and maintenance costs that will raise the unit cost nor expected longer lifetime that will reduce the unit cost. According to the author's fieldwork in China's coal-fired power plants, continuous emissions monitoring system (CEMS) costs about 500,000 RMB/set around 2010. China has been publishing hourly data from CEMSs in key polluting sources. With an expected lifetime of approximately 5 to 10 years, the unit cost would also be about US$1 to US$2 per published data point. Screening often requires multiple observations. OCO-2 has a 16-day ground-track repeat cycle to result in about 23 repeated observations per year for one pixel, or at a cost of roughly US$50 per year. CEMSs in China could provide more than 8,000 hourly observations per year and have an annual cost of about US$8,000 to US$16,000. Accordingly, the costs of an average screening technology are assumed to be in the range of several hundred U.S. dollars per year for one polluting source. In contrast, compliance monitoring by environmental inspectors is cheaper to set up but more expensive to operate. For example, China in 2015 at the central level had 542 employees for environmental inspections (Figure 3.1), with a total cost of 63.5 million RMB (~US$10.2 million in 2015 exchange rate, or US$18,800/person-year; Ministry of Environmental Protection, 2016b). Accordingly, the average cost of one inspection was about 0.074 inspector-year/inspection / 70.5% × US$18,800/person-year, or US$2,000/inspection. In the current scenario, the unit cost of a screening technology (r_s) is then assumed to be one order of magnitude cheaper than that of screening technology, or equivalently 0.0074 inspector-year per screening round.

The compliance monitoring accuracy of one technology is hard to exactly measure, because only data on observed compliance and noncompliance are available but not those on absolute truth. Furthermore, the dichotomy of compliance and noncompliance does not measure the severity of noncompliance, while more severe cases, due to their stronger signal-to-noise ratios, tend to be easier to catch.

In theory, the screening strategy would only work when the noncompliance rate in the high-risk group is higher than that in the low-risk group. The more different their noncompliance rates between these groups gap are, the better the screening strategy will be. In the current scenario, K_{1s}, K_{1d}, K_{2s} and K_{2d} are assumed to be 90%, 99%, 70% and 90%, respectively.

7 Environmental technology and industry[1]

1 Goal-centered SO_2 mitigation path

Besides other critical measures, pollution mitigation often involves facilities such as those installed in coal-fired power plants to remove sulfur oxide (SO_2), nitrogen oxide (NO_x), particles, mercury and carbon dioxide (CO_2), together with renewable-energy facilities for reducing coal consumption such as wind turbines and solar panels and hybrid and electric vehicles. Two major factors determine how rapidly a country could utilize these facilities for pollution mitigation. First, there must be a strong demand for their rapid deployment and normal operation, as examined in detail in Chapters 5 and 6. Second, if the demand is put in place, enough supply capacity should be established to meet the demand. A developing country could take the latecomer's advantage to utilize the supply capacity in developed countries. However, because of China's sheer size, the rest of the world might not be able to accommodate its huge demand. With constrained supply capacity but significantly greater demand, the international price of pollution control facilities could rise sharply, and this would discourage their utilization and slow the pollution mitigation process. Rapid pollution mitigation in China relies greatly on the rapid establishment of a domestic industry.

SO_2 mitigation achieved rapid progress over the past two decades from low starting positions. On the supply side, in the late 1990s, China had few domestic firms and barely any commercialized technologies. The Chinese markets were dominated by foreign firms and foreign technologies. After a decade, a large number of firms entered the market to meet the newly emerged huge demand for SO_2 scrubbers to even drive down prices substantially.

As an illustration of the differences between goal-centered and rule-based governance, the progressive paths in China and the United States have been dramatically different in reaching the wide deployment of SO_2 scrubbers in coal-fired power plants and their normal operation of high SO_2 removal rates (Figure 7.1). From the very beginning, the normal operation of SO_2 scrubbers in the United States with rule-based governance has been achieved while the progress went mainly through the deployment dimension. In contrast, China deployed SO_2 scrubbers with poor operation in the early stage and then proceeded simultaneously in the dimensions of deployment and operation until the technical limits of

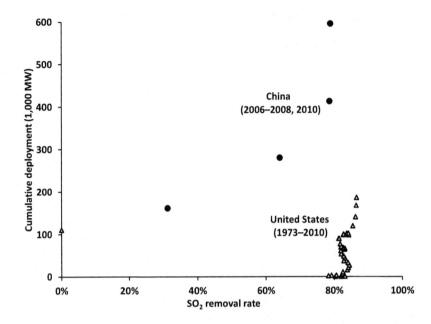

Figure 7.1 The progressive paths on the deployment and operation of SO₂ scrubbers in China and the United States

Source: Lefohn et al. (1999); Xu (2011b); Ministry of Environmental Protection (2011b, 2008–2012); EIA (1986–2006, 2007–2011); Xu (2013).

SO₂ removal rates were roughly reached. Accordingly, the requirements on the quality of SO₂ scrubbers were initially low in the Chinese market and became increasingly higher only later, while in the U.S. market, quality was important from the beginning.

In China, under goal-centered governance, at the early stage of deployment with few SO₂ scrubbers and incapable policy implementation, more SO₂ mitigation would be achieved if the focus were on further deployment rather than on operational improvement. With more and more SO₂ scrubbers in place, any improvement in the operation of the growing stock would lead to a greater reduction in SO₂ emissions. For achieving their SO₂ mitigation goals, the rational choice of the Chinese central and local governments led to a path in which initial progress was made mainly in deploying more SO₂ scrubbers, and it was only afterward that their level of operation caught up.

Implementing policies on the deployment and normal operation of SO₂ scrubbers require different amounts of resources for compliance monitoring. On one hand, the compliance monitoring on the physical existence of SO₂ scrubbers is straightforward and the huge sizes – for example, an absorbing tower is generally several meters in diameter and tens of meters high – make them easily visible. The one-by-one inspection indicates that the corresponding compliance monitoring costs for each SO$_2$ scrubber do not greatly differ, regardless of how many have been deployed. On the other hand, the compliance monitoring on the installation is just a onetime event, but when in operation they demand significantly more resources on a day-by-day basis. A well-functioning environmental compliance monitoring system has significant initial costs of establishment. A significant proportion of additional costs for monitoring one more SO$_2$ scrubber are largely borne by the polluting firms because they are responsible for installing their own monitoring equipment. For policy enforcers, the compliance monitoring costs have a great economy of scale and they increase relatively modestly with wider deployment of SO$_2$ scrubbers.

The political resistance against the deployment and against the normal operation of SO$_2$ scrubbers also differs. The normal operation and maintenance (O&M) costs are significantly higher than the annualized capital costs, especially for SO$_2$ scrubbers with compromised quality (Xu, 2011b). Data on the capital costs of SO$_2$ scrubbers were retrieved from two sources to report a dramatic reduction together with an expanding domestic market of SO$_2$ scrubbers (Figure 7.2). From February to August 2006, China's Association of Environmental Protection Industries surveyed SO$_2$ scrubber projects in operation or under construction at the end of 2005 (Xu et al., 2006). One hundred thirteen projects (223 coal-fired power units) with

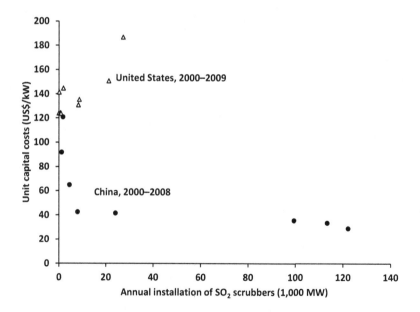

Figure 7.2 Annual average unit capital costs of SO$_2$ scrubbers in China and the United States

Source: Xu et al. (2006); EIA (2012–2013); Ministry of Environmental Protection (2008–2012); Xu (2013).

Note: China's average unit capital costs refer to limestone-gypsum wet scrubbers. Annual average exchange rates were used for currency conversion. Data from 2000 to the peak year of deployment are shown.

a total capacity of 83,850 MW applied limestone-gypsum wet scrubber technology and had cost information available. Data on projects using other technologies are much less continuous to provide longitudinal insights. They had already been or were expected to be in operation over the period from 2002 to 2008. Their expected time in operation could partly reflect when the contracts were signed and accordingly the then market situation. Furthermore, the author's interviews provided an independent source to cross-check the survey data and to shed light on their more recent changes. Data for the United States came from the U.S. Energy Information Administration (EIA; 2012–2013).

Quality has a great impact on the capital costs of SO_2 scrubbers. For example, SO_2 scrubbers in Hong Kong's two coal-fired power plants were contracted with firms from mainland China, and the unit capital costs were three to four times those of similar projects in mainland China, although still at about half of the comparable costs in the United States. Hong Kong's SO_2 scrubbers require high-quality equipment, engineering and construction and enough redundancy, and they take about twice the amount of time from contract to completion. Beyond higher labor costs, the higher price in Hong Kong above "The China Price" could be mainly explained as a quality premium.

Considering the reduction of capital costs in the Chinese market (Figure 7.2), the investment for one more SO_2 scrubber would decrease to indicate that the political resistance dwindles when many SO_2 scrubbers had been deployed. The O&M costs for each SO_2 scrubber varied less along the deployment dimension because of the necessary consumption of electricity, limestone, and water (Table 6.1). More SO_2 scrubbers led to greater overall O&M costs, and this increased the overall political resistance. However, installing SO_2 scrubbers without normal operation wasted financial resources, and it conflicted with environmental policies. The associated political pressure for each SO_2 scrubber from the civil society, despite its underdeveloped status in China, and from within the government increased when more SO_2 scrubbers were deployed to make the problem more visible. The overall net political resistance against the normal operation of existing SO_2 scrubbers could increase at the very early stage of deployment and then shrink when more SO_2 scrubbers are in place.

Given China's then poor record of implementing environmental policies, the evolving quality requirements contributed to goal attainment with a rapid path that could be theoretically understood. The Chinese government can make a certain amount of effort to work for pollution mitigation with two choices, either to deploy more pollution control facilities or to enhance the operational performance of the existing stock. The goal is to maximize the impacts of efforts on pollution mitigation at every step. After a certain amount of pollution control facilities have been deployed, the net political resistance against the deployment of one more facility and against the enhancement of operational performance by 1% could be roughly taken as unchanged with the level of deployment. Accordingly, a given amount of effort could either raise the deployment rate by α (in the two cases of SO_2 scrubbers, the unit is megawatts, MW) or the performance of existing facilities by $\beta\%$ (in the SO_2 scrubber case, the unit is percentage points of SO_2 removal rates). The initially deployed facilities have a total capacity of A, and the initial performance is $B\%$. Then the initial pollution mitigation effect of the facilities is roughly proportional to $A \times B\%$. The performance has a technical upper limit, $B^*\%$.

The option of devoting the efforts to the deployment could raise the pollution mitigation effect to $(A + \alpha) \times B\%$, and the other option to work on the operation would have an effect of $A \times (B\% + \beta\%)$. If there is no constraint, a rational decision maker to maximize the impact of his or her efforts will choose the first option when $(A + \alpha) \times B\% > A \times (B\% + \beta\%)$, or when $\dfrac{A+\alpha}{A} > \dfrac{B\%+\beta\%}{B\%}$ or $\dfrac{\alpha}{A} > \dfrac{\beta\%}{B\%}$. The second option will be taken when $\dfrac{\alpha}{A} < \dfrac{\beta\%}{B\%}$, and the two options are no different when $\dfrac{\alpha}{A} = \dfrac{\beta\%}{B\%}$. With the progress on the deployment and operation, the choice could change. This is what goal-centered governance would indicate. If adding one constraint that the choice should prioritize policy enforcement, the progress should be first made to improve the operation. Only when $B\%$ has reached $B^*\%$, more facilities are allowed to be deployed. This could illustrate rule-based governance.

One more constraint could be added to describe the situation on the supply side. As examined below with more details, the goal-centered governance strategy lowers technological barriers of market entry to facilitate the rapid establishment of a large-enough supply capacity, while the rule-based governance strategy would correspond to higher market-entry barriers and discounted supply capacity in the Chinese context. To simplify the model, the supply capacity under rule-based governance is $\eta\%$ less than that in goal-centered governance, and thus, the same amount of efforts could only raise the deployment rate by $\alpha \times (1 - \eta\%)$.

The SO_2 scrubber case is simulated here to exemplify the usefulness of this very simple model. Here are the assumptions of the earlier parameters: (1) A_0: the initial capacity of SO_2 scrubbers, 7,000 MW, equivalent to the level in 2000 (Ministry of Environmental Protection, 2008–2012); (2) $B_0\%$: the initial SO_2 removal rate in coal-fired power plants with SO_2 scrubbers, 31.3%, equivalent to the level in Jiangsu Province in 2006 (Xu, 2011b); (3) $B^*\%$: 79%, the highest SO_2 removal rate China achieved in 2010 (Figure 6.1); (4) $\dfrac{\alpha}{\beta\%}$: $\dfrac{4,500 \text{ MW}}{1\%}$, or the required effort from decision-makers was the same to deploy 4,500 MW of SO_2 scrubbers and to increase the SO_2 removal rate of the existing stock by 1%. The number is assumed to fit China's actual data; (5) $\eta\%$: 50%, assumed to indicate the impacts of higher market-entry barriers in the rule-of-law strategy. As illustrated in Figure 7.3, the projection with the goal-centered governance strategy fits well into China's SO_2 mitigation path for coal-fired power plants with SO_2 scrubbers. If considering no constraint from the supply side, rule-based governance mainly would differ from goal-centered governance at the early stage of progress. However, if considering the potential supply constraints due to higher market-entry barriers, pollution mitigation under rule-based governance would proceed at a much slower pace.

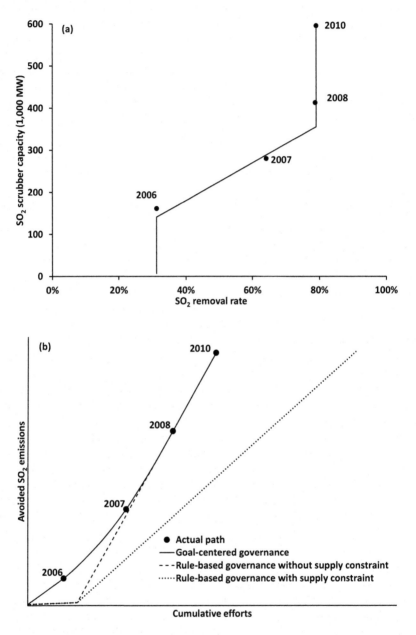

Figure 7.3 Model projection of the SO$_2$ mitigation path in China's coal-fired power plants: (a) deployment and operation of SO$_2$ scrubbers under goal-centered governance (the dots refer to actual data); (b) avoided SO$_2$ emissions under goal-centered and rule-based governance

Source: Xu (2013).

2 Technology licensing under goal-centered SO_2 mitigation path

The international technology market provided opportunities for China's domestic firms to license foreign technologies and to quickly ramp up their technological capabilities, although at a cost. Functioning markets for transferring technologies to developing countries not only are important for their economic development and upgrading along the value chain but also have critical implications for the environment. Due to China's huge and steadily growing emissions, how fast and effective environmentally friendly technologies were adopted was a key determinant for its environmental cleanup. Technology transfer from developed to developing countries has long been recognized as a key measure in addressing CO_2 mitigation (United Nations, 1992). One important method of technology transfer is through technology licensing. With available markets for technologies, a technology owner could choose between licensing its product or directly investing in the client country, and a firm that needs technology could either license in or innovate indigenously (Arora et al., 2001a; Teece, 1988; Arora et al., 2001b). In international negotiation on transferring low-carbon technologies from developed to developing countries, developed countries generally argue for market-based solutions and adequate protection of intellectual property rights (IPR), while developing countries often demand nonmarket solutions at lower than market rates (Ockwell et al., 2010). The differing positions become an obstacle to the agreement of new and effective climate treaties (Ockwell et al., 2010).

Despite unfavorable conditions, the global market for technology has been significant, amounting to about US$35 to US$50 billion in the mid-1990s (Arora et al., 2001b) and roughly US$100 billion in 2002 (Arora and Gambardella, 2010). However, only a small portion – less than one third for the United States – of technological transactions were between unaffiliated organizations and thus true market transactions (Arora and Gambardella, 2010; Saggi, 2002). Most cross-border technology licensing happens among developed countries and that from developed to developing countries is much rarer (Arora and Gambardella, 2010). Product markets in most developing countries are not large enough to attract many potential technology licensors. Developing countries generally lag behind developed countries in human and technological capacities that enable them to effectively absorb licensed foreign technologies and exploit their full value (Metz et al., 2000). Additionally, effective IPR protection could help address the problems of unauthorized use of intellectual property (Gans and Stern, 2010), but developing countries often do not have well-developed systems of IPR protection and thus are placed in relatively disadvantageous positions in creating an attractive market for technology (Strokova, 2010). However, large developing countries like China are able to access foreign low-carbon technologies, although not those at the cutting edge (Ockwell et al., 2010; Lewis, 2007). China's rapid development of many industries had roots partly in the importation of foreign technologies, including, for example, wind turbines (Lewis, 2007), large hydroelectric turbines (Liang, 2001) and high-speed railways (Chan and Aldhaban, 2009).

An especially prominent case was that of SO_2 scrubbers. SO_2 scrubber technologies have been commercially deployed since the mid-1970s, mainly in developed countries. Up until 1998 (expressed in terms of generating capacity of power stations thus equipped), the pace of deployment was about 10 GW per year in the world and 4 GW per year in the United States (Srivastava et al., 2001). Many international firms had established their technological and engineering reputations in this field. China began to significantly deploy SO_2 scrubbers about three decades later than developed countries, with a deployment rate of over 100 GW per year in the 11th Five-Year Plan (Chapter 5). Because of their high SO_2 removal efficiencies – generally over 90% with wet-type technologies – SO_2 scrubbers became the most vital technology in achieving China's goal of a 10% reduction in SO_2 emissions in the 11th Five-Year Plan (2006–2010; Xu, 2011b, 2011c). Among the more than 500 GW of SO_2 scrubbers in China at the end of 2010, more than 90% were installed by Chinese firms using licensed foreign technologies (Ministry of Environmental Protection, 2011a). Major Chinese firms universally licensed foreign technologies and relied heavily on them. Conversely, fewer than 5% were installed by foreign firms or under joint ventures (Ministry of Environmental Protection, 2011a). Domestic firms dominated the market, in spite of their initial lack of proven technologies and experience.

The goal-centered SO_2 mitigation path created three characteristics of China's SO_2 scrubber demand in the early stage. The difficult SO_2 mitigation goals together with China's colossal size required more than 100 GW SO_2 scrubbers annually, which was multiple times as big as the world together had experienced before (Figure 5.12). Their initial poor operation significantly relaxed actual quality requirements (Figure 7.1). The initial one-sided emphasis on the deployment of SO_2 scrubbers indicated that the huge demand for SO_2 scrubbers would be created swiftly from a low level in the 10th Five-Year Plan, which led to stringent time constraints for SO_2 scrubber firms (Figure 5.12). They played key roles in shaping the strategies of domestic technology licensees and foreign technology licensors for tapping into the market.

2.1 The strategy of domestic technology licensees

China's domestic firms as technology licensees could fall into the three following categories: state-owned, university-established and nonstate. "State-owned" firms refer to those controlled by state-owned power corporations, which could have faced less fierce competition to win SO_2 scrubber projects because of their special "internal" relationship. Indigenous SO_2 scrubber technologies had been developed by a few research institutes and universities to directly transfer their human and technological capabilities to state-owned and university-established firms. Nonstate firms could behave differently due to their relative lack of such initial capabilities. In addition, although most of China's major firms relied heavily on licensed technologies, some concentrated on applying their own. China had five large state-owned power corporations at the national level in the late 2010s, four having major SO_2 scrubber firms, and two were selected for interview.

In the available SO_2 scrubbers at the end of 2011 with unit scales not smaller than 100 MW, the two firms had market shares of 11.9% and 3.3%, respectively. Another smaller firm owned by one of the five power corporations was also visited, and its market share was 0.4%. The special relationship with their parent corporations put them in relatively advantageous positions in market competition. Eight firms that had no association with power corporations were interviewed. Their market shares ranged from 0.7% to 6.0%, being 22.8% in total. In addition, two foreign firms and their Chinese representative offices as technology licensors were also interviewed to provide an external perspective.

Domestic firms' decisions to license in SO_2 scrubber technologies were heavily influenced by the three demand characteristics under the goal-centered SO_2 mitigation path. First, the sheer size of China's demand for SO_2 scrubbers challenged the supply capacity. One concern was whether China had enough engineers. This condition was met partly through rapidly training many more university students (Figure 7.4). In 2000, 496,000 undergraduate students graduated from full-time four-year undergraduate programs, including 213,000 in engineering. In 2010, the numbers had grown to 2,591,000 and 813,000, respectively. In 2018, the numbers further climbed to 3,868,000 and 1,269,000, respectively. In 2018, about the same number of undergraduate students (3,665,000) graduated from other full-time programs with shorter study periods of two or three years. The age group, 20 to 24, comprised 5.95% of China's population in 2018, or 16.6 million for each yearly

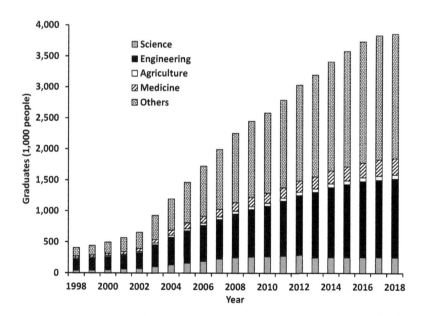

Figure 7.4 Yearly university graduates in China from four-year undergraduate programs by subjects

Source: Ministry of Education (1999–2019).

age (National Bureau of Statistics, 1996–2019). Accordingly, in 2018, about half of China's newly available labor force had a received formal university education. Other part-time or Internet-based undergraduate programs trained another 4.1 million graduates in that year. These enhanced human resources provide a crucial foundation for China's rapid deployment of pollution-removal industrial facilities.

The huge market also helps diminish one concern that licensors might not transfer technologies completely after receiving payments (Arora et al., 2001b). In the case of SO_2 scrubber technology, royalties dominated the revenue stream in technology licensing and effectively deterred such a moral hazard. By way of example, an American firm charged one licensee US$652,118 as the up-front lump-sum fee (Table 7.1): interviews discovered that a license's approximate royalty rate should be 2% of SO_2 scrubber contract values. Between 2004 and 2010, the firm's income from royalties was nearly 40 times as much as the up-front lump-sum fee (the licensee completed 34,900-MW wet SO_2 scrubbers in that period; Ministry of Environmental Protection, 2011a), and the national average contract value was about US$35/kW (Xu et al., 2006)). From another perspective, as demonstrated in the case of a Japanese licensor, a licensee's loss was limited to approximately the up-front lump-sum fee when the technology transfer was not satisfactory. In addition, if a licensor gained a bad reputation, this could limit its future business opportunities in the huge and rapidly growing Chinese market.

Second, the quality requirements for SO_2 scrubbers were initially low. The deployment of SO_2 scrubbers took off around 2002, but the normal operation was improved significantly only in about 2007 (Xu, 2011b; Xu et al., 2009). In the five gap years, many managers of installed SO_2 scrubbers did not plan to operate them normally and cared very little about the quality, while quality was closely associated with the technological advancement of a supply firm. In addition, China's reform in the power sector in 2002 created multiple independent power corporations to

Table 7.1 Up-front lump-sum fees of SO_2 scrubber technology licenses (the Chinese licensees here are all listed on stock markets and the data are from their annual reports)

Chinese licensee	Country origin of the foreign licensor	Lump-sum fee*	Year	Technology type
Wuhan Kaidi	Germany	US$277,304	1998	Dry type
Fujian Longjing	Germany	US$3,989,234	2001	Wet type Circulation fluidized bed
Wuhan Kaidi	United States	US$652,118	2002	Wet type
Zhejiang Feida	United States	US$1,250,000	2002	Wet type
Jiulong Electric	Japan	US$1,126,563	2002	Wet type
Jiulong Electric	Austria	US$1,423,765	2004	Wet type
Insigma Technology	France	US$1,200,000	2004	Wet type

* Exchange rates on December 31, 2010 were used: 1 US$ = 6.62 RMB = 0.75 euro.

encourage competition – this was even though all of these were state-owned. The rapid construction of new power plants strained their available financial resources to create strong incentives to minimize capital investment for each new project, while the poor quality of SO_2 scrubbers could substantially reduce capital costs. Furthermore, the low requirement for quality was strengthened by the largely separate decisions of capital investment and daily operation and by the different incentives of respective decision-makers. Managers of coal-fired power plants should have an incentive to install high-quality SO_2 scrubbers while capital investment was within the authority of the upper levels of management in power corporations. The low requirement for quality and technological advancement substantially lowered the technological market-entry barrier not just for the SO_2 scrubber firms but also along the entire supply chain. In contrast, the quality requirement and technological market-entry barrier in the U.S. market were much higher.

China's regulators also paid attention to the quality requirements, especially with the knowledge of domestic firms' initially unsatisfactory technological statuses. Technologies could come from international transfer or in-house innovation. Various factors could affect the choice of a country or a firm between these two technology strategies. China used to focus almost entirely on in-house innovation under the rule of Chairman Mao when China segregated itself from the world. The "Not Invented Here" syndrome – that internally developed technologies are preferred – was found to be a barrier to technology licensing (Arora and Gambardella, 2010), but it does not seem to be deeply rooted in China in the economic reform era. Secondary innovation based on imported technologies, coupled with original and integrated innovation, had been established as three cornerstones of China's indigenous innovation strategy (State Council, 2006). With regard to the installation of SO_2 scrubbers, China stipulated in tendering documents that established technologies were required. As late as 2005, bidders were clearly asked to specify a foreign technology provider that had installed SO_2 scrubbers of the same or greater scale (Guizhou Qiandong Power Station, 2005). Interviews also confirmed the general requirement for foreign, commercialized technologies in the early years when almost no Chinese firms had any proven experience. This requirement was relaxed only in later years after many firms in the market had completed enough projects.

Third, time was a serious constraint. In the late 1990s and early 2000s, few domestic firms were capable of designing SO_2 scrubbers. The sudden appearance of a huge market led to the creation of many new firms and the reorientation of existing ones from other industries. Because few firms had any prior experience and the market was large enough to accommodate many, most – except those owned by coal-fired power corporations – were placed on a more or less equal footing. Firms would achieve distinction if they could establish engineering and management teams and develop their technological capability faster than others. Another time constraint was the short period from the issue of tendering documents to completion of the bidding process; this typically lasted only one to four weeks. Additionally, the design process could not take more than a few months if the construction was to begin on schedule. Successful firms had to respond quickly and provide acceptable quality.

These time constraints helped push domestic firms toward technology licensing, due to their weak technological foundations. When demand for SO_2 scrubbers started to surge, domestic technologies were generally not able to satisfy the time constraints because of their immaturity. Domestic research and development generated "naked" technologies, to quote the word of one interviewee. Demonstration projects on a commercial scale should be followed by multiple projects to make the technology mature and ready for wide commercial deployment. The commercialization of these "naked" technologies would require at least a few years plus significant financial resources and the willingness of coal-fired power plants to take risks by trying them. The expected short-term peak in China's scrubber market diminished the potential return on investment in indigenous technology. The easy prospect of licensing foreign technologies also reduced the incentive to take risks with indigenous innovations. All the major Chinese firms in the market licensed foreign technologies in order to acquire and substantiate their technological capabilities. No clear difference could be found among state-owned, university-established and nonstate firms. Even the nonstate firm that mainly applied its own technology had to initially license from abroad.

As tacit knowledge cannot be so easily transferred as codified knowledge, know-how played a positive role in establishing a sound market for technology. The contractual acquisition of know-how presents more problems than licensing patents (Arora et al., 2001b). However, in a developing country like China with poor IPR protection, the licensing of patents might be unnecessary in the absence of know-how as the knowledge contained in the patents have already entered the public domain. Chinese firms had generally chosen to legally license, rather than to illegally acquire, SO_2 scrubber technologies. Legal licensing secured a complete package including systematic training, technical documentation and trade secrets in a relatively short timescale, without exposing the licensees to legal disputes. One alternative option was to recruit experts from foreign firms, but the legal risks were not insignificant and the received technologies may not be complete because it would be difficult to recruit an entire team. It would also take much longer for the acquiring firms to comprehend a technology by this means than they would through technology licensing. The associated costs would not be low either, because foreign experts generally had to be paid considerably more than standard Chinese salaries. Furthermore, illegal acquisition did not provide a technological guarantee from a trusted provider, while this guarantee was stipulated by coal-fired power plants in their tendering documents.

China's domestic firms could quickly absorb licensed technologies to meet the time constraints. From as early as the 1970s, China had, through its own research and development on SO_2 scrubbers, built up vital capabilities to establish domestic firms and assimilate imported technology (Shu, 2003). From the mid-1970s to the mid-1980s, China appraised several technologies, although on scales that were at least one or two orders of magnitude smaller than any commercial project. For example, a 300-MW unit corresponds to a flue gas flow rate of about 1,000,000 Nm^3/hour (cubic meter at standard temperature and pressure per hour), while the largest Chinese experiment at the time had a flow rate of 70,000 Nm^3/hour

(Shu, 2003). From the mid-1980s to 2000, foreign technologies were demonstrated on a commercial scale (Gu, 2004; Shu, 2003). In 2000, having resulted in a considerable fund of domestic human and technological capability, foreign technologies were officially recognized as the basis for further development of SO_2 scrubber technologies in China (National Economic and Trade Commission, 2000). China's absorptive capacities were effectively distributed to all major firms including non-state ones through a free labor market of engineers and managers.

Recognizing the constraints of technology licensing such as on expansion beyond China, in the past two decades, China has put a much heavier emphasis on research and development (R&D). In 2000, China had 922,000 full-time equivalent personnel on R&D and this number rapidly grew by 375% to 4.4 million in 2018. R&D expenditures were raised from 0.60% of gross domestic product (GDP) in 1995 to 2.19% in 2018 (Figure 7.5). A much more vibrant market for technology emerged and the transaction value increased from 0.46% of GDP in 1995 to 1.97% in 2018 (Figure 7.5). Together with the rapid growth of China's GDP, the R&D expenditures and technology market transaction values had become 1084% and 1365% greater in 2018 from the levels in 2000 in real terms (Figure 7.5). This R&D boom strengthened China's capacity to absorb foreign technologies and innovate domestic intellectual property. In the category of environmental technology, China's residents and nonresidents were granted 103 and 69 patents, respectively, in 2000 in China's patent filing office, which were about 10% of those in the United States. They grew to 7,459 and 881 patents,

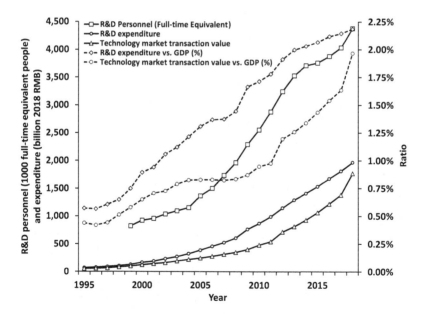

Figure 7.5 R&D personnel, expenditure and market value (in 2018 RMB) in China
Source: National Bureau of Statistics (1996–2019).

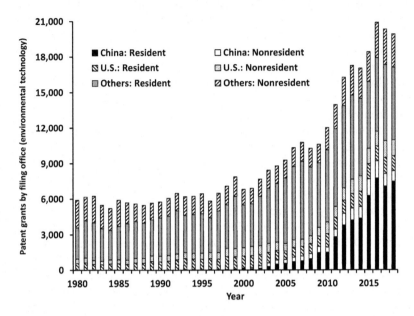

Figure 7.6 Patents on environmental technology by filing office in the world
Source: WIPO (2019).

respectively, in 2018, while the figures in the United States were correspondingly 1,258 and 1,369 patents (Figure 7.6).

2.2 The strategy of foreign technology licensors

The strategy of potential foreign technology licensors was also shaped by the previously mentioned three characteristics of China's SO_2 scrubber demand under a goal-centered SO_2 mitigation path. First, the huge demand for SO_2 scrubbers created profitable business opportunities. Their decision of technology licensing involves the revenue effect (i.e., payments received from licensing) and rent-dissipation effect (i.e., revenue loss due to a new or strengthened competitor in the product market; Arora and Fosfuri, 2003). A stronger revenue effect promotes the decision to license, while a stronger rent-dissipation effect discourages licensing. For major foreign firms that held intellectual property of SO_2 scrubber technologies, the option to do nothing was rarely attractive because of the temptation of the huge emergent Chinese market. The revenue effect was indeed significant. Technology licensing only required a small office in China to monitor licensees and to "service" the partnership. For example, each of the two interviewed American firms had an office in Beijing with about five staff members, whereas their licensees were in charge of contracts worth several hundred million dollars annually. The initial cost in transferring technologies was covered by up-front lump-sum fees paid by licensees (Table 7.1). The commercial success of licensees would result in considerable royalties to the licensor if the contracts were honored. After the know-how and trade secrets were transferred, the intellectual property rights were at risk of misuse or infringement, possibly with the royalties not being fully paid. Despite this, most foreign firms decided to take this risk in order to avoid the much greater risk inherent in direct investment.

After technologies are transferred, one primary concern of technology licensors arose on whether licensees paid royalties honestly. Both licensors and licensees reported in interviews that major Chinese firms were paying royalties regularly. Also, several expiring licenses had been renewed, indicating a good record of royalty payments. As a preventative measure, design software was encrypted and only specially prepared computers could install it with annual reregistration. Several interviewees in the Chinese firms said that, after a few years, they had figured out what was inside the black box but still chose to pay royalties. It was not very difficult to keep track of licensees. The huge size of SO_2 scrubbers often made local news and the Ministry of Environmental Protection annually published details of every SO_2 scrubber and its contractor (Ministry of Environmental Protection, 2011a). Besides, a good partnership with licensors suited the long-term interests of licensees. Technological sophistication had increased step by step in the Chinese SO_2 scrubber market as reflected in the unit scales: the 300-MW scale was dominant before 2005, but after 2006, the 600-MW scale became crucial and then the 1,000-MW scale or greater (Ministry of Environmental Protection, 2014). Every significant increase in scale indicated a new technical advance. Accordingly, the licensing of scrubber technologies was a continuous operation and not a one-off process. Good partnerships, strengthened by honest royalty payments, could also help licensees expand into new markets through future technology licensing. In addition, a partnership may generate business opportunities for both sides. For example, when a large coal-fired power plant in Hong Kong decided to install SO_2 scrubbers, it first approached several international firms, including one from the United States. But the American firm was fully committed in the domestic market and was not willing to take the financial risk of an Engineering, Procurement, and Construction (EPC) project in Hong Kong. Its Chinese licensee was introduced and finally won the contract.

Royalty rates may decrease over time to reduce the costs of honoring licensing contracts. For example, one license divided the ten-year contract period into three phases with declining royalty rates. In several other cases, the royalty rate was renegotiated when competition in the market became much too fierce to significantly shrink the profit margin. Excessively high royalty rates could damage licensees' competitiveness. The final result might be a reduced income from royalties and an increased risk of no payment being made at all. The renegotiation strengthened the partnerships between licensors and licensees and thus worked for the interests of both sides. In one licensing contract signed in 1998, the level of royalties was originally associated with the volume of flue gases. Because China's capital costs of installing SO_2 scrubbers had dropped substantially since then (Figure 7.2), the royalty rate would increase significantly as a percentage of the contract value. Renegotiation took place to lower the royalty rate. The partnership remained strong with both the licensor and the licensee maintaining market success.

Lawsuits, particularly those resolved outside China, were also a deterrent to potential infringement, which maintained the strong revenue effect. For example, Insigma Technology is a Chinese firm listed on the Shanghai Stock Exchange, and it releases information regularly. It signed a technology licensing contract with a French firm in December 2004 (Table 7.1). However, in April 2006, Insigma declared that it would cancel the contract and thereafter stop using the licensed technology. Royalties were paid for six projects in 2005 and 2006 with a total capacity of 7,450 MW (Sina Finance, 2010). The firm later signed a new contract with an Italian firm in September 2006, which was for one year and was to be automatically renewed if no objections were received from either side. The fee for royalties was a fixed sum of €20,000 (US$26,600) for every project regardless of the contract value (Sina Finance, 2010). The French firm later sued Insigma in Singapore (where disputes should be resolved according to the licensing contract). The court made a decision in February 2010 and Insigma was ordered to pay compensation of US$2,085,737 for the loss of royalties in 2005 and US$24,566,684 for the loss afterward (Sina Finance, 2010). The lawsuit may have helped deter other significant licensees from not honoring their licensing contracts.

Second, low-quality requirements and correspondingly low technological market-entry barriers led to active market entry of new firms to contain the rent-dissipation effect for technology licensors. If the downstream operations of a firm are small or the downstream market is in fierce competition, the rent-dissipation effect will be limited and technology licensing becomes more likely (Arora and Gambardella, 2010). Indeed, the Chinese downstream SO_2 scrubber market was newly created and in fierce competition (Ministry of Environmental Protection, 2011a). In addition, market evolution also demonstrated that the rent-dissipation effect should be minimal. Foreign firms tended to lag behind domestic ones in understanding the market's real demand, especially in the early period. Among all the foreign firms, the examined Japanese firm ought to be the best prepared for the Chinese market. It owned more Chinese patents on flue gas desulfurization than any other firm (State Intellectual Property Office, 2010) and, between the late 1980s to 1990s, had won contracts to install China's first-ever commercial wet SO_2 scrubbers (four units of 360-MW capacity; Gu, 2004). However, up to the end of 2010, its technology was only applied to a further 3,300 MW, with the final project in 2006 (Mitsubishi Heavy Industries, 2011). Interviews in China revealed that many foreign firms generally licensed design software together with other know-how in order to enable their Chinese licensees to compete independently, but this Japanese firm was reluctant to hand over design software and wanted to participate more actively. Thus, the technology transfer of know-how was not complete. The decision could have been influenced by the expectedly significant rent-dissipation effect due to potentially high rents as a result of its favorable position in granted patents. However, partly because the relationship made them slower in responding to the market and hampered their competitiveness, its Chinese licensees decided instead to do business with other technology licensors. For example, according to the annual reports from a firm listed on the Shanghai Stock Exchange – Jiulong Electric, the holding firm of Yuanda Environmental Protection

Engineering – although US$1.1 million was paid to the Japanese firm as the up-front lump-sum fee, just two years later it decided to sign another licensing contract with a European firm and gave up the Japanese technology (Table 7.1). Even with the tight control of technology licensing, the Japanese firm earned little profit or rent from the Chinese market, an indication of a small rent-dissipation effect. The existence of many technology licensors diminished the rent-dissipation effect because no single licensor had significant market power.

Third, time constraints discouraged direct participation of foreign technology licensors in the Chinese market. Two interviewed American firms each had a small representative office in Beijing, but their licensing strategies were notably different. They reported that the Chinese government put no restrictions on allowing foreign firms to bid for SO_2 scrubber projects, but many foreign firms did not expect that they would earn significant profits by establishing subsidiaries or joint ventures in China. One major American firm expected the Chinese market to peak for only a few years before it began shrinking; this expectation proved prescient (Figure 5.12). The initial investment of capital and human resources to establish a subsidiary in China would therefore only be of temporary benefit. The firm's past experience in other countries suggested that direct investment could not be freely withdrawn, and accordingly, it was not justified in this particular Chinese market. In addition, the lack of adequate human resources also constrained some foreign firms from choosing direct investment, particularly due to the revived U.S. market for SO_2 scrubbers (U.S. Energy Information Administration, 2011).

2.3 Why technology market can emerge in China?

Even in developed countries – as Gans and Stern argue – an effective market for technology is difficult to establish because it often fails to satisfy the three criteria of effective market design as specified by Roth that successful marketplaces must be "thick, uncongested and safe" (Gans and Stern, 2010; Roth, 2008). The Roth criteria were proposed to fix broken markets or build new ones if they are missing, which could be especially useful for environmental protection as market failure is often the cause. First, an efficient market requires many potential buyers and sellers, or market thickness, to enhance the chances of effective matching. However, many ideas are not independent but reliant on other complementary ideas and assets to achieve their full value, with notable examples in low-carbon technologies (Harvey, 2008). This problem makes the licensing of a single idea less desirable. If the ideas belong to different entities, ineffective coordination could limit the willingness of potential buyers and sellers to participate in the market. Second, the market should overcome Roth's "congestion" criterion, whereby buyers and sellers should be able to negotiate with a number of possible trading partners and have sufficient time to make effective selections. In a congested market, competition is not sufficient and the price does not reach market equilibrium. Because necessary information disclosure for buyers to assess a technology's value might lead to unwanted diffusion, the information is often kept secret between buyers and sellers to constrain open market competition, thus failing the "congestion" criterion.

Third, market transactions should be "safe"; that is, conducted in good faith and with safeguards that allow the expression of real intention and information and result in mutual satisfaction. A drawback on this point is that, after licensors have disclosed information, licensees might be able to exploit it independently, without signing licensing contracts, creating issues over misuse of intellectual property.

The Chinese market for SO_2 scrubber technologies satisfied all three Roth criteria. Key contributing factors could include China's large market size, the maturity of available technologies and goal-centered governance. First, because the size of the Chinese market for SO_2 scrubbers as a downstream market for the technologies is far greater than any other country, major foreign SO_2 scrubber firms, as potential licensors, could hardly overlook the potential business opportunities. The large market and low technological barriers facilitated by technology licensing have created many domestic firms as potential licensees. Multiple sellers from the United States, Europe and Japan actively licensed out their technologies (Xu et al., 2009, 2006). In addition, in the Chinese market up to 2010, 16 firms – all Chinese – had completed at least 10 GW of SO_2 scrubbers, all using licensed-in foreign technologies (Xu et al., 2006; Ministry of Environmental Protection, 2011a). The three types of Chinese firms – state-owned, university-established and nonstate – did not show significantly different behavior in the market for technology. Fierce competition drove down costs and diminished expected profit from direct investment, but revenue from technology licensing was significant. The rent-dissipation effect was overwhelmed by the revenue effect of technology licensing, which accordingly became a dominant choice of foreign firms. As a large country, China has a strong capacity to absorb new technology due to its previous R&D, and this capacity was effectively distributed to all three types of firms through a free labor market. Licensors and licensees held multiple bilateral negotiations simultaneously to help solve the market congestion problem. Furthermore, the safety of technology licensing also benefited from China's large market size. As a result of the large market, there were significant revenues from royalties that encouraged licensors to transfer complete packages of technologies. The market for SO_2 scrubbers at every unit scale was substantial and the unit scales escalated over time to require continuous technological support from licensors. Such dynamism favored long-term partnerships between licensors and licensees for their mutual benefit and fostered honest royalty payments.

Second, the maturity of SO_2 scrubber technologies played a crucial facilitating role. After several decades of commercial deployment in developed countries, many firms had acquired complete technology packages. Personal and corporate expertise, or know-how as tacit knowledge, was a vital part of the technology package. Acquiring knowhow raised costs and contracting problems, but given the inadequate standard of IPR protection in China, technology licensing became necessary in order to acquire complete packages of technologies. Many foreign firms had become independent technology holders, and a potential licensee only needed to negotiate with one licensor for a complete technology package. When deciding whether to license out technologies or set up direct subsidiaries in developing countries or even just do nothing, firms from developed countries needed to compare the expected profits of each market option. The dominant business reality in the market was technology licensing. For a potential licensee, the technology could either be developed internally or acquired externally. Favorable conditions created the demand for foreign technologies in the Chinese market.

The Chinese market also met the second Roth criterion on the lack of congestion. The maturity and wide deployment of SO_2 scrubbing technologies also enabled a fairly accurate estimation of the technology's value to facilitate market transactions. Interviews revealed that, although the negotiation of technology licensing was generally bilateral, without disclosing information to third parties, licensors and licensees often negotiated with several entities on the other side at the same time for most suitable licensing contracts. IPR protection is recognized as a key means to ensure market safety and satisfy the third Roth criterion (Gans and Stern, 2010). As examined earlier, know-how and credible threat of lawsuits ensured the general satisfaction of this criterion. The disclosure of the necessary information for value assessment in negotiations caused fewer problems because knowhow could not be easily acquired.

The existence of many potential licensees enabled licensors to design their strategies to maximize profit. At least three clear strategies emerged among three licensors. A major American firm licensed to only two Chinese firms and built up long-term partnerships through full technical support. One license was restricted to the licensee's home province for a certain period and the other covered the whole of mainland China. The licensees had a near monopoly to use the specific technology in their assigned market territories. Another significant American firm had about eight licensees in China; the strategy was to increase the market share of its technology as well as its royalties, but the licensees were still selected so as to prevent unqualified ones from ruining the technology's reputation. In addition, as mentioned earlier, a Japanese firm licensed its technology to a few Chinese firms but, unlike the two American firms, refused to transfer design software. The two American firms had their technologies widely applied but the Japanese technology was abandoned without much deployment. From the perspective of the level of royalties, the two American strategies were clear winners.

An effective market for cutting-edge technologies is understandably more difficult to establish. It is probable that not many organizations have acquired intellectual property as potential licensors. The value of a particular cutting-edge technology is harder to assess and the accumulation of know-how may still be in progress with a consequently high price of the final product which will limit its deployment. These unfavorable conditions discourage the emergence of potential licensees. Information disclosure to facilitate licensing will also raise more concerns on the part of technology owners. As a result, the Roth criteria of effective market design will be harder to meet for cutting-edge than for mature technologies.

Third, goal-centered governance resulted in a path of SO_2 mitigation to significantly lower market-entry barriers for domestic firms. The previous two factors are mainly given, while governance strategy could be more deliberately taken. For developing countries that have not established a sound rule of law and strong domestic industries for pollution removal, goal-centered governance may induce a feasible path for improvement. In order to meet time constraints and technological requirements, major Chinese firms universally licensed in foreign technologies to quickly build technological strength. In the early period, China had not established a system to well implement environmental policies and thus many SO_2 scrubbers were not operating normally. For meeting governmental regulations, coal-fired power plants chose to install the cheapest SO_2 scrubbers but did not expect to run them. For domestic firms that had no technological advantages, this initially low but escalating requirements on the quality of SO_2 scrubbers provided helpful stepping-stones to enter the market.

The utilization of wind energy followed a comparable path under goal-centered governance, which also helped to lower market-entry barriers for the establishment of a domestic wind turbine industry. Similar to the SO_2 mitigation case, the initial stage of wind energy development also focused more on the deployment to follow the goal-driven demand. In China's 11th Five-Year Plan for Renewable Energy Development, the major goal for wind electricity referred to generation capacity whereas actual electricity generation served as a supplementary goal (NDRC, 2008). One average kilowatt-hour of wind capacity consistently generated much less electricity in a year in China than in the United States, and this partly indicated poorer operating conditions in China (Figure 7.7). When the deployment of wind turbines became sufficiently wide, the Chinese government started to pay more attention to their operation. Problems in the quality and operation of wind turbines emerged with their deployment to threaten not just wind

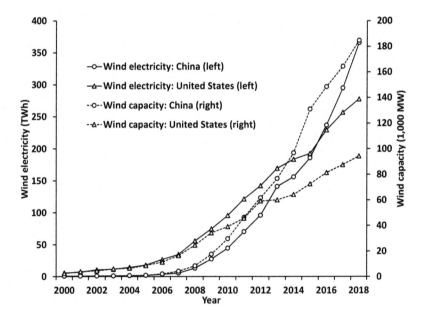

Figure 7.7 Wind energy development in China and the United States

Source: BP (2019).

electricity generation but, more important, also the safety of the electric grid and to push for greater focus and higher requirements (SERC, 2011). In 2010, the National Energy Administration published a plan to enact 247 technical standards for wind energy development, including several which were already in force (National Energy Administration, 2010). Lower technological market-entry barriers played a positive role to encourage new firms. In 2006, the Chinese market had 12 firms that supplied wind turbines, and the number rose to 29 in 2012 (Shi, 2007; China Wind Energy Association, 2012). Many component suppliers along the supply chain also actively entered the market (Chinese Wind Energy Equipment Association, 2011). Compared to wind turbine manufacturers, market-entry barriers were even lower and the technologies were less complex for component suppliers, and this resulted in fiercer competition and thinner profit margins.

Furthermore, unlike SO_2 scrubber firms, the Chinese firms in the wind industry licensed their technologies from a very different category of foreign firms. Foreign licensors of SO_2 scrubber technologies were generally major firms that were closely involved in the downstream business of installing SO_2 scrubbers (Xu, 2011a). In contrast, major foreign wind turbine manufacturers were largely reluctant to license technologies to Chinese firms, and most foreign licensors were design firms or small manufacturers that focused more on upstream technological development. This phenomenon is explained in the theory of markets for technology as the rational choice based on the respective industrial structure (Arora and Gambardella, 2010). The good-enough quality, lower price and no geographic constraints of technology licenses made the Chinese domestic wind industry potentially competitive.

The market for technology might also work for other large developing countries, such as India. They may also have potentially large markets through which to spawn many domestic operators and fierce competition. Many other low-carbon and pollution-control technologies have been commercialized with much know-how. A caveat is that these large developing countries may not necessarily always have large domestic markets for pollution mitigation. These are partly determined by government policies and not just by the overall sizes of their economies. Their abilities to take on board foreign technologies might not be consistently strong. However, there is great potential for large developing countries to make use of markets for technology to build their industrial prowess with mature technologies. Goal-centered governance may provide more feasible pathways for domestic industries in these developing countries to take roots and further grow from weak starting points.

3 Environmental industry under goal-centered SO_2 mitigation path

3.1 Market entry and competition

Considering both firms that pollute the environment and others that provide pollution removal facilities, the impacts of the goal-centered SO_2 mitigation path in China may not be straightforward. On one hand, although empirical studies generated mixed results on the "pollution haven hypothesis" in the Chinese context

(Levinson and Taylor, 2008; He, 2006; Shen, 2008), its key root cause – poor environmental regulation, including weak policies and poor enforcement – is argued to potentially benefit polluting firms for not acting on, delaying or comply only partially with pollution control (Harney, 2008). In China, policies on environmental protection and business standards were recognized by polluting firms as less important barriers to market entry (Niu et al., 2012). Relative to the Euro4 fuel quality standards, the poorer Euro2 standards in China could reduce costs by 1.1 and 1.9 U.S. cents per gallon for gasoline and diesel, respectively (Liu et al., 2008). The cost burden also acts as a political and regulatory hurdle to bring polluting firms under full compliance. On the other hand, from the perspective of supplying pollutant removal facilities, weak regulation could lower market-entry barriers to encourage competition, innovation and the establishment of industrial capacities for pollution control (Stigler, 1971; Dean and Brown, 1995).

Two important barriers on the supply side could slow down the deployment of SO_2 scrubbers in China. No existing supply capacity could meet the unprecedented peak demand of over 100 GW a year (Figure 5.12). The capital costs of about US\$65 to 90/kW (Figure 7.2) were initially too high, being over 10% of the costs of building new coal-fired power plants (SERC, 2006). If the large labor force and industrial base in China could be effectively mobilized for the deployment of SO_2 scrubbers, the supply capacity would not have a major problem in meeting the rapidly growing demand. The lack of significant restrictions on foreign direct investment indicates that both foreign and domestic firms could tap into the labor force.

The huge Chinese market can easily accommodate many SO_2 scrubber firms without losing economies of scale. Whether the supply potential could be released depends on whether existing firms could expand their capacity and (more importantly) whether new firms could emerge. Although the U.S. market had only about ten firms, and with new firms rarely entering, the Chinese market had over 60 firms – almost all of which were newly established, most being domestic but some being foreign – thereby indicating much lower market-entry barriers (Figure 7.8). In the past decade, the annually added capacity of SO_2 scrubbers increased significantly both in China and the United States, but the evolution of unit capital costs showed a rapid cost reduction in China and a cost spike in the United States (Figure 7.2). In China, the rapidly rising demand triggered intensive market entry to create fierce competition followed by a cost reduction whereas competition in the United States was rather limited, and this constrained the expansion of the supply capacity. When the demand for SO_2 scrubbers grew, the price was pushed up.

As discussed earlier, domestic firms did not have technological advantages, especially in the early period. Nevertheless, because of the existence of many potential licensors in the technology market, no foreign firm was able to prevent others from licensing technologies to China. Technologies therefore could not be used as a barrier to exclude Chinese firms from competing. The crowded market enabled fierce competition not just for providing SO_2 scrubbers. Competition also took place between foreign firms for licensing to especially promising Chinese firms that were expected to win many projects and return significant revenues

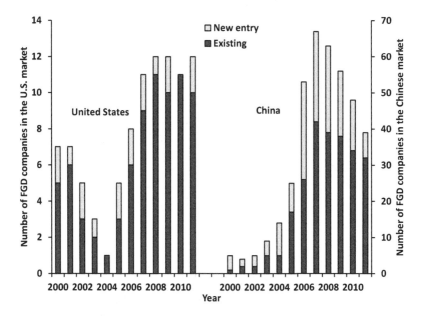

Figure 7.8 Firms in the Chinese and U.S. markets installing 100-MW-scale or greater SO₂ scrubbers

Source: Ministry of Environmental Protection (2008–2012); EIA (2007–2011); Xu (2013).

Note: "Existing": firms have been in the market in the past. "New entry": firms entering the market for the first time. The U.S. numbers use the left axis, and the Chinese numbers use the right axis.

from royalties. Those potential licensees were mainly established by coal-fired power producers. Interviews showed that financial payments were the most critical aspect of negotiating licenses, although other aspects were also important, such as the suitability of technologies and the scope of licenses. The willingness to accept lower up-front lump-sum fees and lower royalty rates made a licensor more competitive. After the significant variance of early contracts, the up-front lump-sum fee stabilized to be about US$1.2 million for wet scrubbers (Table 7.1).

3.2 International competitiveness of China's SO₂ scrubber industry

Due to specific features in various environmental fields, goal-centered governance may present very different impacts on different environmental industries. One significant difference is on the international competitiveness of China's SO₂ scrubber and wind turbine industries, as could clearly be seen from the reaction of the United States to China's rising industrial prowess. Over the same period as China's rapid growth was taking place, the United States also witnessed significantly wider deployment. From 2004 to 2010, its SO₂ scrubber capacity grew from 100 GW to 181 MW and its wind capacity from 6.8 GW to 40.3 GW (EIA,

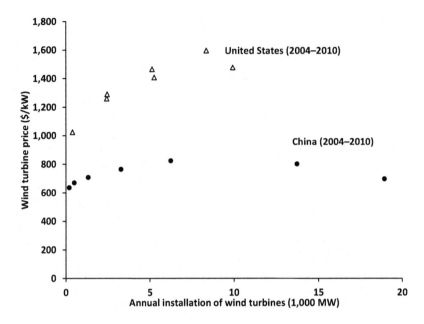

Figure 7.9 Average prices of wind turbines in China and the United States

Source: IEA and ERI (2011); Wiser and Bolinger (2012); BP (2019); Xu (2013).

2012–2013). "The China Price" was a critical reason for trade disputes between China and the United States. In 2010, the price tag of SO_2 scrubbers in China was about US$20/kW as revealed in the author's fieldwork, whereas in the United States, it was US$206/kW (EIA, 2012–2013). For wind turbines, the average price in 2010 was US$700/kW in China and US$1,460/kW in the United States (Figure 7.9). However, China's SO_2 scrubbers barely made any news in trade disputes between the two countries while those of wind turbines were highly visible (Cooper, September 28, 2012). From another perspective, the Chinese SO_2 scrubber industry did not contribute to international SO_2 mitigation whereas its wind industry strengthened the global CO_2 mitigation capability.

Despite the success in building up the supply capacity and achieving cost reduction, China's large SO_2 scrubber industry did not become competitive in the international market as indicated by the nearly tenfold price difference in the segregated Chinese and U.S. markets (Figure 7.2). Many SO_2 scrubbers were of low quality, and this increased the operation and maintenance costs and shortened their lifetimes. Although the delayed improvement of the operation of SO_2 scrubbers was critical for lowering the initial quality requirement and technological barriers to market entry, after 2007 when the normal operation of SO_2 scrubbers was largely expected, the prices stayed low. The gap between 2002 and 2007 was too long and China was trapped in a low-quality bottom. The huge quality premium presented serious financial challenges to power corporations. In addition, the quality of SO_2 scrubbers was quite opaque to investors, and only the SO_2 scrubber firms had the best knowledge of the product. In the five gap years, a race to the bottom had pushed the quality and price of SO_2 scrubbers to reach a minimum and stable level. Because no SO_2 scrubber firm had established a reputation for quality, any significant price increase would put the firm in a disadvantageous position in competition. Even when China started to allow BOT (Build, Operate, Transfer) contracts for SO_2 scrubbers to better integrate the decisions of capital investment and daily operation (NDRC and SEPA, 2007), the trap remained a difficult one to escape from. Another important reason for the segregation of the Chinese and U.S. SO_2 scrubber markets lay in the restriction of technology licensors. Almost every major Chinese SO_2 scrubber firm licensed and relied on foreign technologies that felt themselves constrained in the Chinese market (Xu, 2011a). Even projects in Hong Kong required special permission from technology licensors.

However, the lower market-entry barrier at the early stage of wind energy development was still much higher compared to that of SO_2 scrubbers. Although costs were much lower in China than in the United States, a race to the bottom on quality and price did not happen and the price of China's wind turbines remained stable (Figure 7.9). The operational requirement never dropped to a bottom as in the SO_2 scrubber case. One critical reason lay in their different regulatory foundations. Although the enforcement capacity for the deployment and operation of SO_2 scrubbers could be built on the existing regulatory system, the weak environmental policy enforcement indicated that such a system had not been satisfactorily established in China. In comparison, the compliance monitoring system for wind electricity delivery had been largely established despite wind energy being a new energy type for electricity supply. Furthermore, because electricity generation has direct and significant economic benefits to local governments, the political will for greater demand and better management was much stronger than in the case of SO_2 scrubbers. Because the poor operation or quality of wind turbines would affect wind electricity generation and thus the revenue, investors in wind farms value quality substantially more than those investing in SO_2 scrubbers.

Despite the highly visible trade disputes between China and the United States, the actual trade in wind turbines was minimal. In 2011, the total capacity of exported wind turbines was equivalent to only 1.3% of that installed domestically (China Wind Energy Association, 2012). Although four Chinese wind turbine manufacturers had been ranked among the largest ten in the world, unlike the other six as regional or global suppliers, they remained largely domestic (Li et al., 2011). Besides other influential factors, one important reason could be the quality gap that made the Chinese wind turbines fail to reach the technological market-entry barriers in developed countries. However, the Chinese wind industry could have a promising future. If the price difference between China and the United States were taken as the upper limit of the quality premium or the depth of the quality trap, the wind industry would be much more likely to escape the trap than the SO_2 scrubber industry.

As demonstrated in the two comparative case studies, the depth of the low-quality trap could be determined by how long the operational improvement of pollution control facilities is delayed. The delay should be long enough for the domestic supply capacity to become established but short enough to prevent a race to the bottom on quality and price. Another influential factor on the depth of the trap is how strong the initial enforcement capacity is. Because electricity generation corresponds to much stronger enforcement capacity than the mitigation of conventional pollutants, China could have a better chance to build internationally competitive industries for renewable energy that generally has to be converted into electricity. Low market-entry barriers for quality and technological advancement are a key factor to make the Chinese market and industrial development vibrant. In the later upgrading, China could focus more on raising the corresponding requirements but on keeping other barriers low to minimize the negative impacts of such enhancement.

4 Inter-goal coordination under goal-centered governance

China's Five-Year Plans feature multiple goals in several fields, including economy, social development, environmental protection and resource conservation. Goals on economic growth rates are always the first one in the goal table in each Five-Year Plan, while they have been listed as "expecting" since the 11th Five-Year Plan when goals were first differentiated between "expecting" and "binding" (National People's Congress, 2001, 2006, 2011, 2016, 1996). Although goals on environmental protection have been gaining importance and become "binding," the relationship between economic development and environmental protection is still crucial to profoundly affect the sustainability of the environmental political will and the achievement of environmental goals. One pivotal concern is how to coordinate various goals for maximizing their potential synergies and minimizing conflicts. SO_2 mitigation and economic development have two-way impacts. First, SO_2 mitigation is one constraint for economic development. Energy consumption and economic growth are fundamental drivers of SO_2 emissions, whose mitigation thus reversely becomes a limiting factor. Second, SO_2 mitigation also relies on the emergence and development of a pollution removal industry to feasibly provide the technological means of SO_2 mitigation, which could create new jobs and economic opportunities.

Over the past four decades, central economic planning has also gradually shifted toward decentralized market evolution. Various local governments are also actively competing with each other in establishing local industries that can serve the huge national market. One key feature of the four-decade economic reform has been the gradual peeling of constraints on the market. The state-owned sector has been generally retreating and those remaining ones are more profit-driven than like governmental agencies. China's economic reform has created many markets from a negligible basis after the Cultural Revolution and greatly enhanced the importance of the markets. The boundary between the state and the market has also become clearer.

China's SO_2 mitigation path as examined earlier surely has contributed to its SO_2 mitigation goals. At the same time, new economic opportunities emerged and were generally seized, which should also have facilitated the advancement of economic goals. In comparison with rule-based governance, goal-centered governance has resulted in much lower requirements on inter-goal coordination. Local governments in China are the primary, decentralized entities to bear the responsibilities and incentives for achieving both environmental and economic goals. They can have greater flexibility in adapting their policies and actions to take the best advantage of changing situations.

These goals are also crucial indicators of how the Chinese central government balances between environmental protection and economic development. When economic goals were emphasized while environmental goals were not, local governments primarily focused on achieving economic goals. These goals are not fully coordinated but generally are independently implemented in a bottom-up manner. They do not demand centrally planned coordination either, as shown previously in China's surprising emergence of the SO_2 scrubber industry. They will seek appropriate ways for balancing how they achieve both goals. Decentralized policies and market evolution may utilize unexpected opportunities and circumvent unexpected difficulties in a much better way than any intelligent central planner can foresee in advance. Goal-centered governance thus can better maximize synergies and minimize conflicts among various goals and government tasks.

Note

1 Adapted with permission from Xu, Y. 2011. China's functioning market for sulfur dioxide scrubbing technologies. *Environmental Science & Technology,* 45, 9161–9167. Copyright (2011) American Chemical Society; and Xu, Y. 2013. Comparative advantage strategy for rapid pollution mitigation in China. *Environmental Science & Technology,* 47, 9596–9603. Copyright (2013) American Chemical Society. Much has been revised and expanded on.

References

Arora, A. & Fosfuri, A. 2003. Licensing the market for technology. *Journal of Economic Behavior & Organization,* 52, 277–295.

Arora, A., Fosfuri, A. & Gambardella, A. 2001a. Markets for technology and their implications for corporate strategy. *Industrial and Corporate Change,* 10, 419–451.

Arora, A., Fosfuri, A. & Gambardella, A. 2001b. *Markets for technology: The economics of innovation and corporate strategy.* Cambridge, MA: MIT Press.

Arora, A. & Gambardella, A. 2010. Ideas for rent: An overview of markets for technology. *Industrial and Corporate Change,* 19, 775–803.

BP. 2019. *Statistical review of world energy* [Online]. Available: www.bp.com/en/global/corporate/energy-economics/statisticalreview-of-world-energy.html.

Chan, L. & Aldhaban, F. 2019. *Technology transfer to China: With case studies in the high-speed rail industry.* PICMET 2009 Proceedings, Portland, OR, August 2–6, 2858–2867.

China Wind Energy Association. 2012. *Statistics on China's wind turbine installation in 2011.* Beijing, China: China Wind Energy Association.

Chinese Wind Energy Equipment Association. 2011. *Chinese wind turbine generator system selection manual (edition 2011)*. Beijing, China: China Wind Energy Association.

Cooper, H. 2012. Obama orders Chinese company to end investment at sites near Drone Base. *New York Times*, September 28.

Dean, T. J. & Brown, R. L. 1995. Pollution regulation as a barrier to new firm entry – initial evidence and implications for future-research. *Academy of Management Journal*, 38, 288–303.

EIA. 1986–2006. *Annual steam-electric plant operation and design data (Form EIA-767)*. Washington, DC: U.S. Department of Energy.

EIA. 2007–2011. *Electric generator report data (form EIA-860)*. Washington, DC: U.S. Department of Energy.

EIA. 2012–2013. *Electric power annual 2010–2011*. Washington, DC: U.S. Department of Energy.

Gans, J. S. & Stern, S. 2010. Is there a market for ideas? *Industrial and Corporate Change*, 19, 805–837.

Gu, X. 2004. *A summary of installing SO_2 scrubbers in Luohuang power plant and the impact on the society and environment*. Annual conference of the Chinese Association of Science, Hainan, China.

Guizhou Qiandong Power Station. 2005. *Tendering document for the flue gas desulfurization island*. Zhenyuan, Guizhou [Online]. Available: www.in-en.com/power/html/power-2006200604145584.html [Accessed January 27, 2011].

Harney, A. 2008. *The China price: The true cost of Chinese competitive advantage*. New York: Penguin Press.

Harvey, I. 2008. *Intellectual property rights: The catalyst to deliver low carbon technologies*. Breaking the Climate Deadlock, Briefing Paper. London: The Climate Group.

He, J. 2006. Pollution haven hypothesis and environmental impacts of foreign direct investment: The case of industrial emission of sulfur dioxide (SO(2)) in Chinese provinces. *Ecological Economics*, 60, 228–245.

IEA & ERI. 2011. *China wind energy development roadmap 2050*. Paris, France: IEA, ERI.

Lefohn, A. S., Husar, J. D. & Husar, R. B. 1999. Estimating historical anthropogenic global sulfur emission patterns for the period 1850–1990. *Atmospheric Environment*, 33, 3435–3444.

Levinson, A. & Taylor, M. S. 2008. Unmasking the pollution haven effect. *International Economic Review*, 49, 223–254.

Lewis, J. I. 2007. Technology acquisition and innovation in the developing world: Wind turbine development in China and India. *Studies in Comparative International Development*, 42, 208–232.

Li, J., Cai, F., Tang, W., Xie, H., Gao, H., Ma, L., Chang, Y. & Dong, L. 2011. *China wind power outlook 2011*. Beijing, China: China Environmental Science Press.

Liang, W. 2001. Power equipment of the gigantic three Gorges project. *Proceedings of the Fifth International Conference on Electrical Machines and Systems*, 1, 676–678.

Liu, H. A., He, K. B., He, D. Q., Fu, L. X., Zhou, Y., Walsh, M. P. & Blumberg, K. O. 2008. Analysis of the impacts of fuel sulfur on vehicle emissions in China. *Fuel*, 87, 3147–3154.

Metz, B., Turkson, J. K. & Intergovernmental Panel on Climate Change, Working Group III. 2000. *Methodological and technological issues in technology transfer*. Cambridge and New York: Cambridge University Press.

Ministry of Education. 1999–2019. *Education statistics yearbook of China*. Beijing, China: Ministry of Education.

Ministry of Environmental Protection. 2008–2012. *The list of China's SO$_2$ scrubbers in coal-fired power plants*. Beijing, China: Ministry of Environmental Protection.

Ministry of Environmental Protection. 2011a. *China's capacities of water treatment plants, SO$_2$ scrubbers and NO$_x$ removal systems at coal power plants*. Beijing, China: Ministry of Environmental Protection.

Ministry of Environmental Protection. 2011b. *Statistical data on the environment*. Beijing, China: Ministry of Environmental Protection.

Ministry of Environmental Protection. 2014. *The list of China's SO$_2$ scrubbers in coal-fired power plants*. Beijing, China: Ministry of Environmental Protection.

Mitsubishi Heavy Industries. 2011. *Delivery record* [Online]. Available: www.mhi.co.jp/en/products/pdf/delivery_record.pdf [Accessed January 27, 2011].

National Bureau of Statistics. 1996–2019. *China statistical yearbook*. Beijing, China: China Statistics Press.

National Economic and Trade Commission. 2000. *Key planning points on flue gas desulfurization technologies and their localization (2000–2010)*. Beijing, China: National Economic and Trade Commission.

National Energy Administration. 2010. *A set of standards for wind energy in the energy sector*. Beijing, China: National Energy Administration.

National People's Congress. 1996. *Outlines of the 9th five-year plan and long-term goals in 2010 for economic and social development of the people's republic of China*. Beijing, China: The 4th Conference of the 10th National People's Congress.

National People's Congress. 2001. *The outline of national 10th five-year plan on economic and social developments*. Beijing, China: The 4th Conference of the 9th National People's Congress.

National People's Congress. 2006. *The outline of the national 11th five-year plan on economic and social development*. Beijing, China: The 4th Conference of the 10th National People's Congress.

National People's Congress. 2011. *The outline of the national 12th five-year plan on economic and social development*. Beijing, China: The 4th Conference of the 10th National People's Congress.

National People's Congress. 2016. *The outline of the 13th five-year plan on economic and social development*. Beijing, China: The 4th Conference of the 10th National People's Congress.

NDRC. 2008. *The 11th five-year plan on renewable energy*. Beijing, China: NDRC.

NDRC & SEPA. 2007. *Working plan on experimenting BOT management of flue gas desulfurization in coal power plants*. Beijing, China: State Environmental Protection Administration, NDRC.

Niu, Y., Dong, L. C. & Chen, R. 2012. Market entry barriers in China. *Journal of Business Research*, 65, 68–76.

Ockwell, D. G., Haum, R., Mallett, A. & Watson, J. 2010. Intellectual property rights and low carbon technology transfer: Conflicting discourses of diffusion and development. *Global Environmental Change-Human and Policy Dimensions*, 20, 729–738.

Roth, A. E. 2008. What have we learned from market design? *Economic Journal*, 118, 285–310.

Saggi, K. 2002. Trade, foreign direct investment, and international technology transfer: A survey. *World Bank Research Observer*, 17, 191–235.

SERC. 2006. *Capital costs of electric projects completed in the 10th five-year plan*. Beijing, China: SERC.

SERC. 2011. *Supervision report on wind electricity safety*. Beijing, China: SERC.

Shen, J. 2008. Trade liberalization and environmental degradation in China. *Applied Economics*, 40, 997–1004.

Shi, P. 2007. *Statistics on China's wind turbine installation in 2006*. Beijing, China: China Wind Energy Association.

Shu, H. 2003. SO_2 emission control for coal fired power plant. *Electrical Equipment*, 4, 4–8.

Sina Finance. 2010. *Information on Insigma Technology Co., Ltd.* [Online]. Available: http://finance.sina.com.cn [Accessed February 9, 2011].

Srivastava, R. K., Jozewicz, W. & Singer, C. 2001. SO_2 scrubbing technologies: A review. *Environmental Progress*, 20, 219–228.

State Council. 2006. *The outline of national science and technology development plan in the middle and long term*. Beijing, China: State Council.

State Intellectual Property Office. 2010. *Database of patents granted in China*. Beijing, China [Online]. Available: www.sipo.gov.cn/sipo2008/zljs/ [Accessed July 2, 2010].

Stigler, G. J. 1971. The theory of economic regulation. *The Bell Journal of Economics and Management Science*, 2, 3–21.

Strokova, V. 2010. *International property rights index – 2010 report*. Washington, DC: Americans for Tax Reform Foundation; Property Rights Alliance.

Teece, D. J. 1988. Capturing value from technological innovation – integration, strategic partnering, and licensing decisions. *Interfaces*, 18, 46–61.

United Nations. 1992. *United nations framework convention on climate change*. New York: United Nations.

U.S. Energy Information Administration. 2011. *Official energy statistics from the U.S. government*. Washington, DC [Online]. Available: www.eia.doe.gov/ [Accessed May 18, 2011].

WIPO. 2019. *WIPO statistics database* [Online]. Available: https://www.wipo.int/edocs/pubdocs/en/wipo_pub_943_2019.pdf.

Wiser, R. & Bolinger, M. 2012. *2011 wind technologies market report*. Washington, DC: U.S. Department of Energy.

Xu, F., Yi, B., Zhuang, D., Yang, M., Yan, J. & Yan, Z. 2006. *Survey report on the construction and operation of SO2 scrubbers at coal power plants in the 10th five-year plan*. Beijing, China: The 4th Conference on Flue Gas Desulfurization Technologies.

Xu, Y. 2011a. China's functioning market for sulfur dioxide scrubbing technologies. *Environmental Science and Technology*, 45, 9161–9167.

Xu, Y. 2011b. Improvements in the operation of SO_2 scrubbers in China's coal power plants. *Environmental Science & Technology*, 45, 380–385.

Xu, Y. 2011c. The use of a goal for SO_2 mitigation planning and management in China's 11th five-year plan. *Journal of Environmental Planning and Management*, 54, 769–783.

Xu, Y. 2013. Comparative advantage strategy for rapid pollution mitigation in China. *Environmental Science & Technology*, 47, 9596–9603.

Xu, Y., Williams, R. H. & Socolow, R. H. 2009. China's rapid deployment of SO_2 scrubbers. *Energy & Environmental Science*, 459–465.

8 Goal-centered governance

1 Alternative governance models

China is experiencing very serious environmental damage. Nevertheless, the country in the past decade has achieved probably the fastest sulfur dioxide (SO_2) mitigation pace for a large country. Significant progress has been made to clean up air and water. Its energy system has been gaining momentum to transition away from coal and toward renewables. With the economy more than 30 times bigger, SO_2 emissions within one decade dropped to a level that was seen only before the economic reform era began in the late 1970s. Strong political will was formed to increasingly prioritize environmental protection among governmental affairs. The entire Chinese government across the central, provincial, municipality and county levels has been much better mobilized and committed. Policies are constantly enacted by various central and local authorities. The conventional poor policy implementation has been more effectively addressed and rapidly evolving to gain greater efficiency. In the coal-fired power sector, China managed to achieve essentially universal coverage of SO_2 scrubbers. More important, the original nonoperation of SO_2 scrubbers was also reversed to reach high SO_2 removal rates. On the other hand, China established the largest SO_2 scrubber industry, which provided employment and economic outputs. However, two decades ago at the early stage of China's SO_2 mitigation, few domestic firms existed with barely any domestic commercialized technologies. Although China has not been widely recognized by developed countries as a market economy, new firms were actively formed and swarmed into the new market to seek profitable opportunities. China's inadequate protection of intellectual property rights did not seem to have prevented widespread market-based technology licensing from firms in developed countries. Despite numerous problems, China can claim great success in SO_2 mitigation in the past two decades. These different components of environmental governance must work together to witness a favorable outcome. This book assesses the outcome and, most important, aims to explain the trajectory.

Conventional wisdom can easily explain China's environmental crises but has serious difficulties in understanding the cleanup process. Democracy and the rule of law are believed to be crucial contributors to forming strong political will and enabling the means to achieve pollution mitigation. However, China is not a democracy and often ranked much behind developed countries in the rule-of-law index. Accordingly, we expect that China's rapid economic growth will result in environmental crises and unacceptably high SO_2 emissions, but the later, even faster SO_2 mitigation is surprising because it defies the original expectations. China has not been fundamentally changed from the perspective of democracy and rule of law. The Chinese Communist Party is still the ruling political party in China. Governmental officials at various levels are still appointed but not democratically elected. Although certain progress has been made, Chinese society is still far from reaching the similar rule-based status as developed countries.

In one common conventional impression, the Chinese government is authoritarian and highly centralized with forceful central planning. Accordingly, in this theory, China's environmental cleanup in the past two decades would be explained from the perspective of central planning. The central government might have designed the trajectory and its unchallenged authority could then implement such a design. This logic goes that the Chinese government does not have the checks and balances as in those democratic, developed countries, which enables China's central planners to design an optimized path with good coordination among various policy makers and implementers. When few domestic firms existed, the Chinese government did not require the good operation of SO_2 scrubbers to enable low technological market-entry barriers, provide and localize necessary supply capacities and reduce costs of SO_2 mitigation. When many firms have been well established in the market, effluent emission standards and other regulations were made more stringent with better implementation for more effective SO_2 mitigation. These newly emerged environmental industries provide economic opportunities and cushion the negative impacts of stringent environmental protection on economic growth.

However, this explanation must assume that China's central planners were extremely intelligent and well informed, but little evidence shows that such high-quality central planning has ever existed. As a developing country, China's data collection system is less advanced than that in developed countries, especially two decades ago, to provide adequate data support for central planning. China's complexity and scale also make such high-level central planning intelligence impossible to achieve. The extreme centralization under Chairman Mao resulted in social, political and economic chaos with disastrous consequences. It is hardly convincing that central planning can lead to either rapid SO_2 mitigation amid momentous economic growth or the establishment of a large SO_2 scrubber industry.

Furthermore, the rule-based environmental governance that accounts for the trajectories in developed countries can also experience difficulties if applied to provide a primary explanation. As indicated in the World Bank's governance indicators as well as in general impression, China's performance has not been remarkable. China is still unable to make rules as important as developed countries prevalently do for environmental governance. In addition, under rule-based governance, although individual entities make their own decisions based on the rules, the rules are often centrally enacted by legislatures and/or courts as laws and the executive branch as regulations. Even if the rule of law is well established in a society, whether rule-based governance can produce good outcomes depends on the quality of rulemaking. Rule-based governance alone is not a guarantee of a good outcome. Poorly designed rules and effective implementation may turn out to be undesirable, while policy making in China has not gained a decent reputation on its soundness, and consultation has also been much less thorough than that in developed countries. For example, before 1997, market speculation was a serious crime in China that was written into the Criminal Law. The intention was to maintain the order of a planned economy.

This book provides a different account of China's environmental cleanup. China today has abandoned the Soviet-style central planning that was featured in the first three decades of the People's Republic under the leadership of Chairman Mao. However, rule-based governance has not been well established. New laws and policies take a considerable amount of time to form and settle. For example, the Civil Code had just been enacted in May 2020 after many decades of gradual formation. Instead, a new governance strategy has been tried and gradually become mature, with various goals taking the central stage. This goal-centered governance model is a mixture of centralization and decentralization to explain China's SO_2 mitigation trajectory much better than the central planning approach or rule-based governance can.

2 Goal-centered governance

Readings of China are polarized, especially when China becomes bigger and more influential. One side profoundly denounces China and accuses the country of being messy, of not being a democracy, of having a rubber-stamp legislature and of being authoritarian without adequate respect to the rule of law. The Chinese government has been heavily criticized for breaking many rules that are highly valued in liberal democracies, such as those related to political liberty. Freedom of speech and civil society are constrained. Rising income inequality and privileges of the wealthy and the powerful add social tensions. However, another side supports the Chinese government as they see many positive outcomes in China's development. Together with rapid and sustained economic growth, the social welfare system has been expanded dramatically to widen health care coverage even in rural communities, increase retirement pension and alleviate poverty. The Chinese people can now enjoy living standards that were unimaginable one generation ago. They can largely choose where to live, work or travel as well as what to buy and sell. A great majority of the population has received significant returns of the economic development, although the distribution is uneven. Both views on China seem to have strong evidence to validate their claims. Then how can we understand China with these two sharply polarized readings? Are they connected? How China may further reform to embrace a better future?

For evidence-based researchers, the negative views on China could be mainly about rules and their implementation, while the positive views could be primarily shaped by outcomes. Although not all arguments on either side are sound, both views can find enough evidence to back them up. SO_2 mitigation, or environmental protection in general, is one government affair that exemplified such situations. The rapid mitigation was surprising but has been verified from multiple independent data sources, including external satellite data. Although active policy making and effective implementation were pivotal for achieving SO_2 mitigation goals, many policies failed or were not implemented well. Initially, a large fleet of SO_2 scrubbers were built but not normally operating. In any understanding of China's governance, a theoretical explanation should be able to accommodate both sides but not ignore the evidence of the other side. Furthermore, how are the two sides connected? In China's case, does the favorable outcome have to be accompanied by numerous policy blunders? If the rules were required to be well designed and implementable before putting into practice, would that affect the favorable outcomes?

This book explains China's puzzles into a goal-centered governance model. As this book has examined in individual chapters on China's SO_2 mitigation, goal-centered governance has two foci, including goals and policies. Goals direct policies and policies achieve goals. Rule-based governance also has such two foci, but goals become secondary. The decisions in governance are mainly about enacting rules that are expected to be genuinely implemented. Fewer policies (or regulations and laws) are enacted and the policy making might be more centralized, but they tend to be more carefully drafted. The outcome is an implicit product of such rules but not in the form of explicit, binding goals.

The goal-centered governance model can be understood from its organization mechanisms, features and applicability.

2.1 Organization mechanisms

China has two hands in environmental governance, one visible and the other invisible. SO_2 mitigation and environmental cleanup were achieved when the two hands cooperated. As a visible hand, the top leadership sets up prioritized goals with neither full-fledged deliberation nor stringent requirements on the path selection. The path results from bottom-up efforts of decentralized stakeholders as directed by an invisible hand of governance. The invisible hand of the market has been widely recognized and utilized. Rational market participants maximize their self-interests or profits, while this decentralized process also leads to the maximization of a society's overall economic interest. Goal-centered governance could resemble and enable such an invisible hand to guide the central and local governments toward goal attainment. When their self-interests are served with various incentives for goal attainment, the overall goal will be achieved to satisfy society's overall interest. If more stringent goals are enacted, the incentives should also be strengthened. In order to finally achieve environmental cleanup, environmental goals must be prioritized with increasing stringency over a long period. If goals are changed, the invisible hand will direct the system away from the original goals and toward new ones.

As illustrated in Figure 8.1, goal-centered governance comprises three pillars: centralized goal setting, decentralized goal attainment and decentralized policy

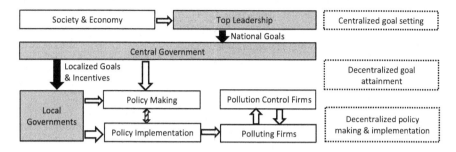

Figure 8.1 An illustration of the goal-centered governance model

making and implementation. First, the process for setting up goals of nationwide priority is highly centralized. The top leadership, with the Political Bureau of the Chinese Communist Party and its Standing Committee at the core, is in charge of supplying the country with goals as they deem crucial, especially in Five-Year Plans. The relationship among different goals could be balanced at this stage. Some goals could be prioritized that correspond to higher ratings in the performance assessment of local leaders. In the case of SO_2 mitigation, the Chinese top leadership did generally respond to what society wants, although the process was not democratic. The goals on SO_2 mitigation and environmental protection were revised more stringent when such demand escalated.

Second, for decentralized goal attainment, national goals are distributed to provincial governments and then lower-level local governments, as in the case of SO_2 mitigation and environmental protection goals. These individualized, quantitative goals guide the efforts of local governments and related ministries. Strong enough incentives are put into place to reward goal attainment and punish failures. Because China's local leaders are appointed but not elected, their jobs are explicitly linked to the performance of achieving various goals with different priorities. The Chinese Communist Party's organization plays a crucial role in establishing such a crucial personnel relationship between the central and provincial governments and their further subsidiaries. In addition, the central government receives much greater revenues than it spends, while the situation for local governments is generally the opposite: to demand a significant fiscal transfer from the central government. If local governments failed their individual goals, their leaders would face grim opportunities of promotion and could even be removed. Those who outperform others are distinguished for better promotion opportunities.

Third, policy making and implementation are heavily decentralized. With the responsibility of achieving goals, local governments have sufficient flexibility, authority and capacity for policy making and especially implementation, while the central government is especially weak in policy implementation. Requirements are significantly lowered on the quality of policy making, the optimal choice of policy instrument and coordination among policies. As a developing country, China has not acquired enough strengths from these perspectives despite continuous improvement. The weak rule of law indicates that the system neither requires nor ensures their genuine implementation. Policies compete with each other and evolve with implementation selection.

2.2 Features

Under goal-centered governance, several key features could emerge.

First, not all goals are important and prioritized goals are few. The mobilization of the entire Chinese government, from central to local levels, depends on the credible incentives for their goal attainment performance. Any additional goal could dilute the effectiveness of existing ones. Accordingly, the number of nationally prioritized goals should be constrained, while provincial governments and central ministries may have their second-tier goals with lower priorities. Governmental efforts are highly concentrated on those goals of high priority, while in areas with lesser or no goals, the performance could be significantly compromised.

Second, policy making is active and each makes an incremental contribution to goal attainment. Local governments and central ministries are mandated to achieve their individualized goals. The incentives are mainly associated with the goals' attainment, while any mistakes in policy making and implementation are much more leniently accommodated. Furthermore, they also have great authority and flexibility in policy making, adoption, innovation and learning in the decentralized arrangement. These favorable conditions encourage active policy making, as witnessed in the case of SO_2 mitigation. Because it is local governments but not their environmental protection bureaus that bear the responsibility of achieving goals, they often involve multiple bureaus in making their specialized policies that may contribute to SO_2 mitigation. Ministry of Ecology and Environment, its predecessors and its composing departments, as well as other central ministries, have also been actively trying new policy tools. Unlike the situation in the United States that the Acid Rain Program in the Clean Air Act Amendments (1990) and its previous versions may claim a lion's share of credits, China does not feature any pivotal policy of similar importance for SO_2 mitigation, while SO_2 mitigation goals were achieved through numerous policies and each contributed a small and accumulative share.

Third, more policy failures exist and policy implementation is selective. These may be seen as the necessary costs of the goal-centered governance model, especially when China is still in the process of strengthening its policy-making quality and policy implementation effectiveness. Policies in China may fail from multiple perspectives. The design itself may be less mature and flawed. Decentralized policy making indicates that not all policy makers, especially those in local governments, have adequate intellectual support. Policy implementation may have unexpectedly high obstacles from various interest groups or weak enforcement capacity. To ensure the faithful implementation of individual policies is only a secondary priority for local governments. When good implementation of a certain policy contributes significantly to goals, more efforts will be directed to this issue.

For SO_2 mitigation, those policies on installing SO_2 scrubbers were first targeted in implementation, while their operation was only made a priority later when the significant and growing fleet of SO_2 scrubbers increased the impacts of such policy on reducing SO_2 emissions. Environmental policy implementation capacity was strengthened, and new environmental compliance monitoring technologies were actively adopted with SO_2 mitigation goals in primary focus.

Fourth, requirements on goal coordination are lower. With impacts on SO_2 mitigation, industrial, energy and environmental policies are enacted generally independently from each other for achieving their specific goals. Various policies for one or multiple goals could have synergies and/or conflicts. In goal-centered governance for SO_2 mitigation, policy coordination largely is not centrally organized. Conflicting policies may not be implemented well to positively contribute to goal attainment, and thus, they would dwindle. Those compatible policies that have synergies will be expanded from local to national levels or adopted from one region to another. In other words, such policy coordination is not achieved primarily through intentional intelligent design but via bottom-up evolution through implementation selection.

Fifth, requirements on information availability and measurability are lower with moral hazards better contained. Policy making and implementation are much more data-intensive than the assessment of goal attainment. Significant uncertainties exist and many potential factors could affect the final outcome, such as in the case of SO_2 mitigation. Because the efforts of local governments are difficult to accurately measure and sometimes hardly observable, local leaders in China may simply pay frequent lip service, emphasize constraints and external factors other than their own efforts but behave differently in reality. Comparison across regions then faces high hurdles to disable effective competition among local governments. However, under goal-centered governance, goals are primarily on those measurable outcome indicators, such as SO_2 emissions and air quality, which significantly reduce the required information. Lip service is much less helpful than actual efforts for achieving goals.

Corresponding to the questions that are raised in the book, the coexistence of favorable outcomes and unfavorable policy pathways is only puzzling because they cannot be properly explained by the rule-based or central planning governance models, while a decent theoretical understanding can be reached with the goal-centered governance model. If the system has a very low tolerance for problems in policy making and implementation, especially for China as a developing country, the favorable outcomes might indeed be seriously compromised. Nevertheless, the costs of policy deficiencies can be reduced when China gradually acquires the capability and capacity for high-quality policy making and effective policy implementation.

2.3 Applicability

Since the Qin dynasty (221–207 BCE) first established centralized rule in China, local governments have always been crucial in Chinese governance to distinguish the importance of the central–local relationship. The vast territory and population, as well as huge regional differences, weaken direct ruling by the emperors or prime ministers who reside in the distant capital. Although China has long been enacting laws and policies in texts, such as those by Shang Yang in a major reform in the 4th century BCE that led to the rise of the Qin Kingdom, the modern sense of the rule of law has never been well established to occupy the central stage of governance.

Corresponding to the organization mechanisms of goal-centered governance, the system may fail under three situations. First, the achievement of governmental goals does not lead to outcomes that the society wants. The supply of goals by the top leadership may have a lag or lead from the demand, but the gap should not be too wide to let the system fail. This concern is closely related to arguments in China's context without democracy. When China was much poorer and the public prioritized economic growth and jobs over environmental protection, environmental goals were ranked much lower than economic goals. When the public started to pay more attention to life quality and clean environment, environmental goals should then be ranked high among governmental affairs. It is not necessary that the goals are exactly identical as what the society desires. For example, the maximization of long-term tax revenues may be compatible with improving the living standard of the public. After the Mongol empire under Genghis Khan occupied North China in early 13th century, one high-ranking official suggested eliminating all Han Chinese and using the land for grazing because Han Chinese's primary economic activities were not raising animals. His goal was for the land to generate more tax revenues. Another key advisor to Genghis Khan, Yelv Chucai, proposed that if the Han Chinese could be left alive to still engage in agriculture and business, they would contribute much more tax. His advice was taken, and the outcome was favorable to both the Mongol court and the people.

Second, the decentralized goal attainment fails. The central government may not be able to impose their prioritized goals onto local governments. A frequent complaint in the Chinese government was that "policies and orders cannot go beyond *Zhongnanhai*." *Zhongnanhai*, or "Central and Southern Seas," is a compound in Beijing where the central government of the People's Republic of China is located. This sentence generally means that the central government cannot smoothly impose their policies and orders onto local governments. Even Chairman Mao complained before the Cultural Revolution that the Beijing municipal government was "penetrable by neither water nor needles." Local governments and central ministries may malfunction or no effective incentives are available to incentivise or force them to work for their assigned goals. A long-lasting question in the Chinese history is the collapse of the Ming dynasty (1368–1644) in early 17th century. Historians pointed out one crucial reason in Emperor Wanli (r. 1572–1620) when he left many key positions vacant and the government could not function (Huang, 1981). In the later decades of the Tang dynasty (618–907), local leaders had exclusive power over military, civil affairs and fiscal revenue. They could also pass their titles to heirs who were chosen by themselves. Essentially, local governments were semi-independent kingdoms, which eventually led to the collapse of the Tang dynasty.

Third, policy making and implementation are overcentralized, and local governments have very limited flexibility or capability in choosing their own paths for achieving goals. One-size-fits-all rules from Beijing may be at a great distance from diverging regional realities to undermine their effectiveness and efficiencies. Active policy making, innovation and learning could be suppressed with overcentralization or when mistakes were much less accommodated. When policy-making authorities, fiscal revenues/expenditures and capable officials are concentrated into the central government, local governments may be too weak to perform their jobs well. Local governments in wealthy regions may experience little difficulty in attracting capable employees or building enough capacity in policy making and implementation for achieving their goals. However, China has significant regional disparity in economic development. If left alone, poor regions would not be able to utilize the policy and technological tools effectively and efficiently.

Goal-centered governance is mainly for new and evolving governmental affairs without well-established policies. In comparison to two decades ago, China has designed, enacted and implemented many policies for SO_2 mitigation and other environmental goals. Many will last to make SO_2 mitigation a routine governmental affair, such as the effluent emission standards of thermal power plants. These tested policies and correspondingly strengthened implementation systems will form an escalating base for the continuous advancement of environmental protection until reaching fundamental solutions. Then goal-centered governance could gradually give way to rule-based governance and other governmental affairs may receive more attention with prioritized goals. In the past two decades, key environmental goals in China's Five-Year Plans have been extended from SO_2 and chemical oxygen demand (COD) in the 11th Five-Year Plan (2006–2010), plus ammonia-nitrogen (NH_3–N) and nitrogen oxide (NO_x) in the 12th Five-Year Plan (2011–2015), plus water quality grade, the Air Quality Index and fine particulate matter ($PM_{2.5}$) concentrations in the 13th Five-Year Plan (2016–2020; National People's Congress, 2011, 2006, 2016). With the continuous progress, it will not be surprising to see that SO_2 mitigation goal removed and an ozone (O_3) goal added in the future, if not in the upcoming 14th Five-Year Plan (2021–2025).

This goal-centered governance has been tested as an effective strategy for China to make rapid advancement from unfavorable situations and to significantly lower key requirements on policy making as in a rule-based governance system. From one perspective, it is an effective and efficient path-finding strategy for China to reach a more sustainable, rule-based future. With new problems continuously emerging, it should and will be the crucial strategy in China's future governance even when China reaches the stage of a developed country.

The goal-centered governance model may be best utilized in countries with the following characteristics: (1) newly prioritized governmental affairs or others with rapid evolution to require continuous focus; (2) developing countries where policies have not been maturely established and policy making has not achieved adequate quality and acquired enough data and intellectual support; (3) being large in scale with genuine necessity of multiple governmental levels and where the central government can impose adequate incentives on local governments to encourage policy innovation, while goal evaluation is largely fair with good data support and rewards are issued based mainly on meritocracy; (4) where the system is more tolerant to mistakes in policy making and implementation and pays primary attention to outcomes and only secondarily on paths; and (5) local governments are capable of policy innovation and resourceful for policy implementation.

Countries in federal systems may not find this governance model applicable because incentives very likely are neither adequately available nor strong enough for the federal government to incentivize state governments. Small countries may not need this governance strategy as the central government is much closer to the society and local governments are not as important as those in large countries. For countries that have established sound rule of law, goal-centered governance may not occupy center stage either because the system is less tolerant of mistakes in policy making and implementation, while active policy innovation may indeed encounter more mistakes and failures. Highly centralized countries in policy making and implementation may constrain such bottom-up efforts as well. This goal-centered governance model is not necessarily inapplicable in democracies, but the application nevertheless may be much constrained if competition across local governments may not have enough impetus and incentives.

Despite the constraints of its applicability, governments at various levels across countries with different institutional and developmental contexts may still be able to draw helpful insights from the goal-centered governance model and explicitly apply goals in organizing their governance. Decentralized policy innovation and competition can be encouraged in countries with sound rule of law, despite various constraints of existing rules.

2.4 Comparison with other theories

This study's development of the goal-centered governance model not only benefits immensely from earlier theoretical explorations but also demonstrates significant differences.

Goal-setting theory in social psychology is one key intellectual source (Latham et al., 2008; Latham and Yukl, 1975; Locke and Latham, 1990, 2002; Locke et al., 1981). The goal-setting theory mainly emphasizes on how goals could enhance task performance of individuals, while goal-centered governance pays primary attention to the performance of local governments, central ministries and other governmental agencies. In addition, the latter has a heavy focus on the flexibility of those decentralized stakeholders in utilizing policies for achieving those goals.

The goal-centered governance model can be regarded as a specific application of pragmatism with clear directions (Alford and Hughes, 2008), while it places goals at the center and makes policies instrumental. The criteria of assessing policies are based on whether they contribute, undermine or have no impacts on goal attainment in actual contexts but not on prior selection. Policy innovation, competition, revision, learning and expansion are common, and specific policies will rarely be unequivocally relied on. This governance model is a theoretical extension of Deng Xiaoping's cat theory. Deng Xiaoping was officially accredited as the "chief architect of China's reform and open-up" by the Chinese Communist Party. However, he did not have a clear long-term blueprint on how China's economic reform should proceed when China just got out of the devastation of the Cultural Revolution, but many doctrines remained strong. As summarized in his famous quote, "regardless of whether the cat is a white cat or a black cat, as long as it can catch mice, it is a good cat." He was less interested in the debate about whether China's economic reform may contain too much capitalism but mainly focused on whether the country can prosper at a faster pace. This strategy was sharply different from Chairman Mao's, under whose leadership China had a stringent restriction on the choice of paths or "cats." Another famous quote could summarize his main idea: "we would rather have socialistic grass than capitalistic grain." This goal-centered governance has clear directions as specified in goals, but the pathfinding is much less constrained.

It also echoes adaptive and polycentric governance to address complexity and uncertainty that emphasize localized solutions (Dietz et al., 2003; Chaffin et al., 2014; Ostrom, 2010). This goal-centered governance emphasizes more on how these solutions could evolve in decentralized and bottom-up manners with motivated local governments under the centralized direction of goals. In comparison to the comparative advantage strategy that advocates good, incremental improvements but not perfect, once-and-for-all solutions to environmental problems (Xu, 2013), this goal-centered governance model is more incorporative to explain in what conditions the comparative advantage strategy will be taken, why it can work and what impacts it may exert on governance. The competition among local governments and other goal bearers borrows the idea from the Tiebout model (Tiebout, 1956), but they are also quite different. The incentives for the competition are not bottom up from local citizens but are top down from imposed goals. For explaining the development of China's environmental industries, the ecological modernization theory may provide an alternative understanding that connects environmental protection with economic modernization (Hajer, 1995; Zhang et al., 2007). The goal-centered governance model, in comparison, explains that the impacts on environmental industries were not intentionally planned, and environmental and economic policies were not deliberately coordinated for new firms in a developing country like China to actively enter the market and grow up.

Incrementalism is another crucial intellectual source to build the goal-centered governance model (Lindblom, 1959; Lindblom, 1979). Neither emphasizes on key, deliberately designed policies with maximized impacts on achieving objectives, but each policy should make incremental but accumulative contributions. However, goal-centered governance does have explicit goals at the center as ends, while policy making is not centralized for finding optimized means. Instead, the incremental improvement was made by decentralized local governments, not by centralized policy makers. Goal-centered governance is compatible with Joseph Stigler's economic theory of regulation (Stigler, 1971). It understands the demand for regulations with an additional key source from goals, while the supply of regulations is decentralized to witness active policy making, innovation and competition.

3 Implications

In the past two centuries, China has tried, voluntarily or involuntarily, many different governance models. When one model was proved ineffective, reforms were attempted, and frequently, revolutions were started. Even under the rule of the Chinese Communist Party since 1949, China has tried sharply different models. Under Chairman Mao, the Chinese government was much more centralized. His goals significantly deviated away from what the society wanted, but no effective checks could counterbalance and prevent his goals from becoming the nation's. The results were disastrous.

Through trial and error and with tremendous costs, China should have found an effective model to govern the vast, complex, developing country with a deep institutional history. The goal-centered governance model has demonstrated its effectiveness and efficiency in fundamentally reversing the rising trend of SO_2 emissions as well as China's multifaceted environmental crises. Nevertheless, the governance model has two potentially highly damaging risks. First, goals may not be formed to satisfy society's demands, like what happened under Chairman Mao. The current focus on environmental protection could have a chance to be disrupted, and thus, the entire governance system would be directed in another direction. Second, overcentralization and low tolerance to policy mistakes may undermine the system's effectiveness and efficiency. Local governments and other governmental agencies may be weakened on the incentives, authorities and capacities of policy making and implementation. One indicator would be whether policy innovation and learning are still active.

A famous quote from Voltaire, a French writer, is that "the perfect is the enemy of the good." The goal-centered governance model is far from being perfect. Even when it achieves great success, the process is full of stumbles, policy deficiencies, unsatisfactory policy implementation and even frequent abuse of governmental authorities. However, as China has tried, alternative governance models may hardly provide better outcomes due to difficulties from uncertainties, complexities and data inadequacy in China's contexts, although they may work well in another country's contexts. Rule-based governance demands high requirements on policy making quality, optimal choice of policy instruments and inter-policy coordination, but these were not China's strengths especially in the early stages of dealing with major issues such as SO_2 mitigation and environmental cleanup. This goal-centered governance is a "good" model but certainly not a "perfect" one due to its numerous weaknesses. Especially for developing countries with many difficulties in policy making and implementation, this proven "good" model provides a promising way to organize governance for achieving what the society deems significantly desirable, while a "perfect" governance model may be unreachable. The pursuit of being perfect should not stop a country from becoming better.

Environmental crises that have accumulated over a few decades cannot be solved within a few years. Efforts should be sustained even when the government changes after elections or leadership reshuffle. In developed countries, the rule-based governance model has been effective to achieve economic prosperity and later sustained reduction of pollution with laws at the center. The gradually formed and tested goal-centered governance model offers a feasible method for China to fundamentally solve environmental degradation problems. The SO_2 mitigation has transcended multiple Five-Year Plans since the 9th Five-Year Plan (1996–2000) under three top leaderships. The demand for environmental quality has grown stronger among the public, and China's top leadership has also been largely supplying national goals to match the demand. It is expected that environmental goals will remain highly prioritized among governmental affairs in China.

Climate change is a much greater environmental problem than any conventional air or water pollution. This goal-centered governance model has also been used in tackling the mitigation of China's greenhouse gas emissions since the 12th Five-Year Plan (2011–2015) when a goal to reduce carbon dioxide (CO_2) intensity by 17% over the five years was first written into the national plan (National People's Congress, 2011). Goal attainment, policy making and implementation have also been heavily decentralized. The market has been actively taking advantage of economic opportunities from CO_2 mitigation to develop, deploy and innovate technologies, such as renewable energy, electric vehicles and energy efficiency. Similar to SO_2 mitigation, CO_2 mitigation has centralized goals, but its actual attainment is largely decentralized. It is expected that this goal-centered governance model will also lead to China's eventual transition of climate mitigation.

References

Alford, J. & Hughes, O. 2008. Public value pragmatism as the next phase of public management. *American Review of Public Administration*, 38, 130–148.

Chaffin, B. C., Gosnell, H. & Cosens, B. A. 2014. A decade of adaptive governance scholarship: Synthesis and future directions. *Ecology and Society*, 19.

Dietz, T., Ostrom, E. & Stern, P. C. 2003. The struggle to govern the commons. *Science*, 302, 1907–1912.

Hajer, M. A. 1995. *The politics of environmental discourse: Ecological modernization and the policy process*. Oxford and New York: Clarendon Press, Oxford University Press.

Huang, R. 1981. *1587, a year of no significance: The Ming dynasty in decline*. New Haven: Yale University Press.

Latham, G. P., Borgogni, L. & Petitta, L. 2008. Goal setting and performance management in the public sector. *International Public Management Journal*, 11, 385–403, 113.

Latham, G. P. & Yukl, G. A. 1975. Review of research on application of goal setting in organizations. *Academy of Management Journal*, 18, 824–845.

Lindblom, C. E. 1959. The science of muddling through. *Public Administration Review*, 19, 79–88.

Lindblom, C. E. 1979. Still muddling, not yet through. *Public Administration Review*, 39, 517–526.

Locke, E. A. & Latham, G. P. 1990. *A theory of goal setting & task performance*. Englewood Cliffs, NJ: Prentice Hall.

Locke, E. A. & Latham, G. P. 2002. Building a practically useful theory of goal setting and task motivation – A 35-year odyssey. *American Psychologist*, 57, 705–717.

Locke, E. A., Saari, L. M., Shaw, K. N. & Latham, G. P. 1981. Goal setting and task-performance – 1969–1980. *Psychological Bulletin*, 90, 125–152.

National People's Congress. 2006. *The outline of the national 11th five-year plan on economic and social development.* Beijing, China: The 4th Conference of the 10th National People's Congress.

National People's Congress. 2011. *The outline of the national 12th five-year plan on economic and social development.* Beijing, China: The 4th Conference of the 10th National People's Congress.

National People's Congress. 2016. *The outline of the 13th five-year plan on economic and social development.* Beijing, China: The 4th Conference of the 10th National People's Congress.

Ostrom, E. 2010. Beyond markets and states: Polycentric governance of complex economic systems. *American Economic Review*, 100, 641–672.

Stigler, G. J. 1971. The theory of economic regulation. *The Bell Journal of Economics and Management Science*, 2, 3–21.

Tiebout, C. M. 1956. A pure theory of local expenditures. *Journal of Political Economy*, 64, 416–424.

Xu, Y. 2013. Comparative advantage strategy for rapid pollution mitigation in China. *Environmental Science & Technology*, 47, 9596–9603.

Zhang, L., Mol, A. P. J. & Sonnenfeld, D. A. 2007. The interpretation of ecological modernisation in China. *Environmental Politics*, 16, 659–668.

Index

Page numbers in *italic* indicate a figure and page numbers in **bold** indicate a table on the corresponding page.

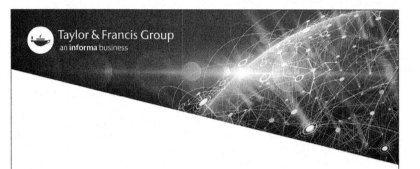

Taylor & Francis Group
an **informa** business

Taylor & Francis eBooks

www.taylorfrancis.com

A single destination for eBooks from Taylor & Francis
with increased functionality and an improved user
experience to meet the needs of our customers.

90,000+ eBooks of award-winning academic content in
Humanities, Social Science, Science, Technology, Engineering,
and Medical written by a global network of editors and authors.

TAYLOR & FRANCIS EBOOKS OFFERS:

A streamlined
experience for
our library
customers

A single point
of discovery
for all of our
eBook content

Improved
search and
discovery of
content at both
book and
chapter level

REQUEST A FREE TRIAL
support@taylorfrancis.com

 Routledge
Taylor & Francis Group

 CRC Press
Taylor & Francis Group

Printed in the United States
By Bookmasters